The SIOP® Model for Teaching Mathematics to English Learners

Jana Echevarría

California State University, Long Beach

MaryEllen Vogt

California State University, Long Beach

Deborah J. Short

Center for Applied Linguistics, Washington, DC
Academic Language Research & Training, Arlington, VA

With contributions by

Araceli Avila

Pearson Education

Melissa Castillo

MelCast Educational Consulting, Phoenix, AZ

Boston   Columbus   Indianapolis   New York   San Francisco   Upper Saddle River
Amsterdam   Cape Town   Dubai   London   Madrid   Milan   Munich   Paris   Montreal   Toronto
Delhi   Mexico City   Sao Paulo   Sydney   Hong Kong   Seoul   Singapore   Taipei   Tokyo

Vice President, Editor-in-Chief: *Aurora Martínez Ramos*
Series Editorial Assistant: *Amy Foley*
Vice President, Marketing and Sales Strategies: *Emily Williams Knight*
Vice President, Director of Marketing: *Quinn Perkson*
Marketing Manager: *Danae April*
Production Editor: *Gregory Erb*
Editorial Production Service: *Kathy Smith*
Manufacturing Buyer: *Megan Cochran*
Electronic Composition: *Nesbitt Graphics, Inc.*
Interior Design: *Nesbitt Graphics, Inc.*
Photo Researcher: *Annie Pickert*
Cover Designer: *Linda Knowles*

For Professional Development resources visit www.pearsonpd.com.

Between the time website information is gathered and then published, it is not unusual for some sites to have closed. Also, the transcription of URLs can result in typographical errors. The publisher would appreciate notification where these errors occur so that they may be corrected in subsequent editions.

Cataloging in Publication data is on file at the Library of Congress.

Photo Credits: p. 1, Ellen Senisi/The Image Works; p. 15, Monika Graff/The Image Works; p. 27, Ellen Senisi/The Image Works; p. 46, Bob Daemmrich/The Image Works; p. 71, Bob Daemmrich Photography; p. 87, Krista Greco/Merrill Education; p. 102, Bob Daemmrich Photography; pp. 116, 129, Bob Daemmrich/The Image Works

Printed in the United States of America

10 9 8 7 6 5 4   BRG 13 12 11 10

www.pearsonhighered.com

ISBN-10: 0-205-62758-7
ISBN-13: 978-0-205-62758-5

Dedication

This book is dedicated to the teachers of mathematics who are committed to the SIOP® Model and who work hard on a daily basis to include language objectives in their math lessons and build academic language. This book is for YOU!

contents

We have written this book in response to the many requests from teachers of mathematics for specific application of the SIOP® Model to their subject. During our nearly 15 years of working with the SIOP® Model, we have learned that both the subject we teach and the students comprising our classes are major considerations in making effective lesson plans. Showing a SIOP® social studies lesson plan to an Algebra teacher, and asking her to "adapt it" has resulted in eye-rolling and under-the-breath comments like, "You've got to be kidding." A similar reaction occurs if we show a physical science lesson video clip to elementary reading teachers and ask them to modify the techniques in their classes. Whatever our subject area, we teachers know what we need and we know what we want.

So, this book is intended specifically for teachers of mathematics, as well as coaches and intervention specialists who work in the area of mathematics. If you teach in grades K–2, 3–5 (or 6), 6–8, or 9–12, you'll find information about teaching math written explicitly for your grade-level cluster. If you are an elementary teacher, obviously you teach multiple subjects and you may want to check out our companion books for teaching English-language arts, social studies, and science within the SIOP® Model.

We offer an important caveat. This book is intended for teachers who have familiarity with the SIOP® Model. Our expectation is that you have read one of the core texts: *Making Content Comprehensible for English Learners: The SIOP® Model* (Echevarría, Vogt, & Short, 2008), or either *Making Content Comprehensible for Elementary English Learners: The SIOP® Model* (Echevarría, Vogt, & Short, 2010) or *Making Content Comprehensible for Secondary English Learners: The SIOP® Model* (Echevarría, Vogt, & Short, 2010). If you have not read one of these books or had substantial and effective professional development in the SIOP® Model, we ask that you save this book for later. Learn the SIOP® Model and then come back and revisit this book when you're ready. It will be more comprehensible and usable if you do so. We want this book to be just what you've been looking for, a resource that will enable you to more effectively teach mathematics to your English learners (and other students). Therefore, the more familiar you are with the philosophy, terminology, concepts, and teaching techniques associated with the SIOP® Model, the better you will be able to use this book. If you would like a refresher on the SIOP® Model, please read Appendix A, Overview of the SIOP® Model. In addition, the eight components and thirty features of the SIOP® Model are listed Appendix B.

The SIOP® Model is the only empirically validated model of sheltered instruction (Echevarria & Short, 2009). Sheltered instruction, or SDAIE (Specially Designed Academic Instruction in English), in general, is a means for making content comprehensible for English learners (ELs) while they are developing English proficiency. The SIOP® Model distinctively calls for teachers to promote academic language development as they promote comprehensible content. SIOP® classrooms may include a mix of native-English speaking students and English learners, or they may include only English learners. This depends on your school and district, the number of ELs you have in your school, and the availability of SIOP®-trained teachers. Whatever your context, what characterizes a sheltered SIOP® classroom is the systematic, consistent, and concurrent focus on teaching both academic content and academic language to English learners.

This book is intended to deepen your understanding of the SIOP® Model and provide more specific teaching ideas, lesson plans, and comprehensive unit plans for teaching ELs in mathematics classes. Our overall goal is to help you master SIOP® lesson and unit planning, enabling you to incorporate the components and features of the SIOP® Model consistently in your classroom.

Organization and Purpose of this Book

This book is for both elementary and secondary teachers. You will read about a wide variety of instructional activities, many of which are effective for any grade level when teaching math. The depth and complexity of a topic will change over the grade levels, of course, but nearly all of the meaningful activities selected to provide practice and application of key concepts and academic language work well with all students, including ELs in grades K–12.

Chapter 1: The Academic Language of Mathematics

Chapter 1 focuses on the academic language that students need to be successful in school. Although it is true that ELs benefit from vocabulary learning strategies that are similar to those used by other students, they generally need more explicit support in vocabulary development (more techniques, for example, that use realia and demonstrations, highlight cognates, and identify words with multiple meanings) and other aspects of academic language. These aspects may be broad-based uses of language such as how to record observations, take notes from a lecture or reference material, or justify orally or in writing the solution steps of a mathematical problem. Academic language also involves more narrow aspects of language, including vocabulary development (such as math terms, and process and function words), and English grammar and usage (such as using transitions properly and writing conditional sentences). Finally, we provide specific examples of the academic language of math culled directly from the pre-K–12 NCTM content standards, so that you can be aware of potential pitfalls for ELs as well as language learning opportunities.

Chapter 2: SIOP® Lesson Planning and Unit Design

In this chapter we focus on SIOP® lesson planning and unit design. We discuss how you can build a week-long math unit that not only covers the regular curriculum but also enables students to make progress with their language development over the course of several days. The chapter begins with some general advice about SIOP® unit planning and the types of decisions teachers must undertake. You will find it interesting to read the questions that guided designing the SIOP® units, their selection of materials, their writing of the content and language objectives, and so forth. We also show and describe two different lesson plan formats that are used in the other chapters of the book.

Chapters 3 and 4: Activities and Techniques for Planning SIOP® Mathematics Lessons

In Chapters 3 and 4, you will find teaching techniques and activities that are effective with English learners and can be used in elementary, middle school, and high school. Organized around the SIOP® components, some of these techniques have been drawn from the *99 Ideas and Activities for Teaching with the SIOP® Model* (Vogt & Echevarría, 2008), and others are suggested by or created by our contributors, SIOP®-trained math teachers Melissa Castillo (elementary grades) and Araceli Avila (secondary grades). Many are not new techniques, but we have specialized them to a particular grade-level cluster: Grades

K–2, 3–5, 6–8, and 9–12. As you read through the lessons in Chapters 3 and 4, pay close attention to how language is embedded into, for example, the mathematical concepts of transformations of parent functions, or geometrical shapes. You may be surprised how much language is involved in teaching mathematical concepts once you focus on it.

Most of the descriptions are followed by a SIOP® lesson plan showing how the techniques are applied in context. Many of the plans have already been field-tested by our contributors and other colleagues.

Chapters 5–8: Sample SIOP® Mathematics Lessons and Units

In these chapters, four units are illustrated, one each for Grades K–2 (Chapter 5), 3–5 (Chapter 6), 6–8 (Chapter 7), and 9–12 (Chapter 8). Our contributors describe their planning process for the SIOP® lessons in each unit presented, discuss objectives and the standards they derive from, the SIOP® techniques and activities they have chosen, and other goals they have. You will find several "think-alouds" and self-directed questions throughout the units through which the writers convey their decision-making process.

For those of you unfamiliar with think-alouds, they are structured models of how successful readers, writers, and learners think about language and learning tasks (Baumann, Jones, & Seifer-Kessel, 1993; Oczkus, 2009). Further, you will notice "Planning Points" comments that clarify and provide additional information, including planning tips. You will also find some of the lessons from Chapters 3 and 4 embedded in the units. Being familiar with the topic and one lesson already should allow you to envision the delivery of the unit more fully. The lesson plans and units also include handouts the teachers will use with students, such as specific graphic organizers and charts, and math worksheets. These materials are found in Appendix C.

Chapter 9: Pulling It All Together

In Chapter 9, we conclude the book with some thoughts, insights, and recommendations from us, as well as from the content specialists and SIOP® experts who served as contributors to this book. We hope this chapter will help you pull it together as you continue your SIOP® journey.

To further assist you in creating a SIOP® classroom, several other resources are also available. These include (in addition to the core texts mentioned previously): *99 Ideas and Activities for Teaching English Learners with the SIOP® Model* (Vogt & Echevarria, 2008); *Implementing the SIOP® Model through Effective Coaching and Professional Development* (Echevarria, Short, & Vogt, 2008); and *The SIOP® Model for Administrators* (Short, Vogt, & Echevarria, 2008).

Acknowledgments

We acknowledge and appreciate the insights and suggestions offered by the educators who reviewed this book. They include Amy Cooper, St. Anthony's Elementary; Lisa deMaagd, Wyoming Elementary; Minda Johnson, St. Anthony's Elementary; and Serena Rosen, Bordentown School District.

To our Allyn & Bacon team we express our gratitude for keeping us focused and on task. To Aurora Martínez Ramos, our incredible editor who never sleeps, we know that it has been through your understanding of the academic and language needs of the English learners in our schools that this series of books has been written on the SIOP® Model. Thanks, Aurora.

● . . . .

xii

We have been most fortunate to have as our contributors for this content area book series eight content specialists and SIOP® experts. Their insights, ideas, lesson plans, and unit plans across the grade-level clusters clearly demonstrate their expertise not only in their content area but also in the SIOP® Model. With deep gratitude we acknowledge the significant contributions of our colleagues: Araceli Avila and Melissa Castillo (Mathematics); Karlin LaPorta and Lisa Mitchener (English-language Arts); John Seidlitz and Robin Liten-Tejada (History-Social Studies); and Hope Phillips and Amy Ditton (Science). Their deep understandings of the SIOP® Model is evident in their techniques, lesson plans, and units, and we thank them for the belief in and commitment to the SIOP® Model.

Finally, we acknowledge and express great appreciation to our families. The SIOP® Model probably wouldn't exist if it hadn't been for your support and encouragement over all these years.

je   mev   djs

Jana Echevarría is a Professor Emerita at California State University, Long Beach. She has taught in elementary, middle and high schools in general education, special education, ESL and bilingual programs. She has lived in Taiwan, Spain and Mexico. Her UCLA doctorate earned her an award from the National Association for Bilingual Education's Outstanding Dissertations Competition. Her research and publications focus on effective instruction for English learners, including those with learning disabilities. Currently, she is Co-Principal Investigator with the Center for Research on the Educational Achievement and Teaching of English Language Learners (CREATE) funded by the U.S. Department of Education, Institute of Education Sciences (IES). In 2005, Dr. Echevarría was selected as Outstanding Professor at CSULB.

MaryEllen Vogt is Distinguished Professor Emerita of Education at California State University, Long Beach. Dr. Vogt has been a classroom teacher, reading specialist, special education specialist, curriculum coordinator, and university teacher educator. She received her doctorate from the University of California, Berkeley, and is a co-author of fourteen books, including *Reading Specialists and Literacy Coaches in the* Real *World* (2nd ed., 2007) and the SIOP® book series. Her research interests include improving comprehension in the content areas, teacher change and development, and content literacy and language acquisition for English learners. Dr. Vogt was inducted into the California Reading Hall of Fame, received her university's Distinguished Faculty Teaching Award, and served as President of the International Reading Association in 2004–2005.

Deborah J. Short is a professional development consultant and a senior research associate at the Center for Applied Linguistics in Washington, DC. She co-developed the SIOP® Model for sheltered instruction and has directed national research studies on English language learners funded by the Carnegie Corporation, the Rockefeller Foundation, and the U.S. Dept. of Education. She recently chaired an expert panel on adolescent ELL literacy. As the director of Academic Language Research & Training, Dr. Short provides professional development on sheltered instruction and academic literacy around the U.S. and abroad. She has numerous publications, including the SIOP® book series and five ESL textbook series for National Geographic/Hampton-Brown. She has taught English as a second/foreign language in New York, California, Virginia, and the Democratic Republic of Congo.

Araceli Avila is an independent consultant and a member of the SIOP®
National Faculty. As a professional developer, her areas of expertise
include bilingual and secondary math education, the SIOP® Model, effective vocabulary instruction, sheltered instruction techniques, instructional
coaching, and adult basic education. She is a former middle and high
school math teacher of English Learners and struggling students and an
adjunct university professor. She is also a certified school counselor in
Texas and a TEXTEAMS trainer. As a former employee of ESC Region 20
in San Antonio, Texas, she supported educators in the areas of Math, ESL,
Bilingual, and Migrant education. She is currently pursuing a doctoral
degree in Educational Leadership.

Melissa Castillo is the lead consultant for MelCast Educational Consulting, working with educators in several states across the country. Areas of
training expertise include the SIOP® Model, academic vocabulary development, coaching, sheltered instruction techniques, and building capacity
for academic success. Her professional experience includes several years
as an elementary teacher in a bilingual and two-way immersion program,
collaborative peer coach, assistant principal, and director of several English Language Learner programs at the primary and secondary levels. As a
doctoral student, her research has focused on effective instructional programs for language minority students, specifically Sheltered English
Immersion.

The Academic Language of Mathematics

Nearly all states continue to struggle to meet the academic targets in math for English learners set by the No Child Left Behind Act. One contributing factor to the difficulty ELs experience is that mathematics is more than just numbers; math education involves terminology and its associated concepts, oral or written instructions on how to complete problems, and the basic language used in a teacher's explanation of a process or concept. For example, students face multiple representations of the same concept or operation (e.g., 20/5 and 20÷5) as well as multiple terms for the same concept or operation

(e.g., 13 different terms are used to signify subtraction). Students must also learn similar terms with different meanings (e.g., *percent* vs. *percentage*) and they must comprehend multiple ways of expressing terms orally (e.g., $(2x + y)/x^2$ can be "two x plus y over x squared" and "the sum of two x and y divided by the square of x") (Hayden & Cuevas, 1990).

So, language plays a large and important role in learning math.

In this chapter, we will define academic language (also referred to as "academic English"), discuss why academic language is challenging for ELs, and offer suggestions for how to effectively teach general academic language as well as the academic language specific to math. Finally, we also include specific academic word lists for the study of mathematics.

What Is Academic Language?

Picture yourself preparing for your first meeting with a financial planner who is going to help you develop an investment portfolio. Memories of hearing or reading about the 1929 stock market crash and the serious economic down turn of 2008 crowd your mind as you gather together your tax returns, payroll stubs, list of assets, and other materials the financial planner has asked you to bring to the meeting. When you arrive at the office, you are introduced to the person who has vast financial knowledge as well as access to additional information about how to establish, maintain, and hopefully profit from a well-managed portfolio. As you seat yourself across the desk from the financial planner, he launches into a detailed explanation of what he thinks you should do. He uses terms and phrases such as:

- Ratio analysis
- Fixed interest securities
- Bulk
- Facultative reinsurance
- Deferred annuity
- Broker-deal
- Convertible

Finally, you recognize a word you know, *convertible*, but you quickly realize this man is not talking about an automobile without a top! You also recognize that the financial planner seems to be speaking a different language, although it certainly sounds like English. How are you going to be able to understand what he is asking you to do when you don't know the language of finance and investment?

We have all had experiences where, as knowledgeable, well-read, educated people, we become lost when we listen to or read about a new and unfamiliar topic. We're often tripped up by the terminology, phrases, and concepts that are unique to the subject matter. When this happens, we most likely become frustrated and disinterested, and we may tune out and give up. Every day, many English learners sit in classrooms where both the topic and the related words and concepts are totally unfamiliar to them. Other ELs may have familiarity with the topic, perhaps even some expertise, but because they don't know the English words, terminology, and phrases—that is, the content-specific academic language— they are also unable to understand what is being taught.

Although definitions in the research literature differ somewhat, there is general agreement that academic language is both generic and content specific. That is, many academic words are used across all content areas (such as *demonstrate, estimate, analyze, summarize, categorize*), while others pertain to specific subject areas (*idioms, characterization, symbolism* for Language Arts; *angle, ratio, dispersion* for Math). It is important to remember that academic language is more than specific content vocabulary words related to particular topics. Rather, academic language represents the entire range of language used in academic settings, including elementary and secondary schools.

When you reflect on the previous examples for Language Arts and Mathematics, you can see that academic language differs considerably from the social, conversational language that is used on the playground, at home, or at cocktail parties (see Figure 1). Social or conversational language is generally more concrete than abstract, and it is usually supported by contextual clues, such as gestures, facial expressions, and body language (Cummins, 1979; 2000; Echevarria & Graves, 2010). To clarify further what academic language is, the following definitions are offered by several educational researchers:

- Academic language is "the language that is used by teachers and students for the purpose of acquiring new knowledge and skills . . . imparting new information, describing abstract ideas, and developing students' conceptual understandings" (Chamot & O'Malley, 1994, p. 40).

- Academic language refers to "word knowledge that makes it possible for students to engage with, produce, and talk about texts that are valued in school" (Flynt & Brozo, 2008, p. 500).

- "Academic English is the language of the classroom, of academic disciplines (science, history, literary analysis) of texts and literature, and of extended, reasoned discourse. It is more abstract and decontextualized than conversational English" (Gersten, Baker, Shanahan, Linan-Thompson, Collins, & Scarcella, 2007, p. 16).

- Academic English "refers to more abstract, complex, and challenging language that will eventually permit you to participate successfully in mainstream classroom instruction. Academic English involves such things as relating an event or a series of events to someone who was not present, being able to make comparisons between alternatives and justify a choice, knowing different forms, and inflections of words and their appropriate use, and possessing and using content-specific vocabulary and modes of expression in different academic disciplines such as mathematics and social studies" (Goldenberg, 2008, p. 9).

FIGURE 1 *The Spectrum of Academic Language*

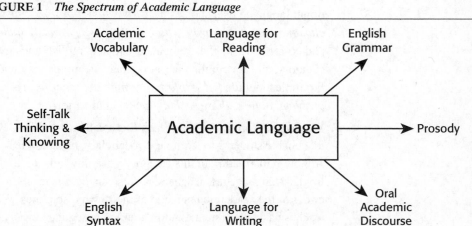

- "Academic language is the set of words, grammar, and organizational strategies used to describe complex ideas, higher-order thinking processes, and abstract concepts" (Zwiers, 2008, p. 20).

It is important to note that some educators suggest that the distinction between conversational and academic language is somewhat arbitrary, and that it is the *situation, community,* or *context* that is either predominantly social or academic (Aukerman, 2007; Bailey, 2007). For purposes of this book, we maintain that academic language is essential for success in school and that it is more challenging to learn than conversational English, especially for students who are acquiring English as a second or additional language. Although knowing conversational language assists students in learning academic language, we must teach English learners (and other students, including native speakers) the "vocabulary, more complex sentence structures, and rhetorical forms not typically encountered in nonacademic settings" (Goldenberg, 2008, p. 13).

How Does Academic Language Fit into the SIOP® Model?

As you know, the SIOP® Model has a dual purpose: to systematically and consistently teach both content and language in every lesson. Content and language objectives not only help focus the teacher throughout a lesson but also (perhaps even more importantly) focus students on what they are to know and be able to do during and after each lesson as related to *both* content knowledge and language development.

A critical aspect of academic language is academic vocabulary. Within the SIOP® Model, we refer to academic vocabulary as having three elements (Echevarria, Vogt, & Short, 2008, p. 59). These include:

1. **Content Words.** These are key vocabulary words, terms, and concepts associated with a particular topic. Key vocabulary typically come from math texts (such as *prime factor, congruence, reliability, subset, centimeter, proof, symbolic representation, histogram*) as well as other components of the curriculum. Obviously, you will need to introduce and teach key content vocabulary when teaching numbers and operations, algebra, geometry, measurement, and data analysis and probability.

2. **Process/Function Words.** These are the words and phrases that have to do with functional language use, such as *how to restate a problem, justify opinions, state a conclusion, work backwards, "state in your own words," summarize, question, interpret*, and so forth. Tasks that students are to accomplish during a lesson also fit into this category, and for English learners, their meanings may need to be taught explicitly. Examples include *list, explain, paraphrase, identify, create, monitor progress of a problem, define, share with a partner*, and so forth.

3. **Words and Word Parts That Teach English Structure.** These are words and word parts that enable students to learn new vocabulary, primarily based on English morphology. Although instruction in this category generally falls under the responsibility of English-language arts teachers, we also encourage teachers of other content areas to be aware of the academic language of their own disciplines. While an English teacher teaches past tense (such as adding an *-ed* to regular verbs) as part of the English-language arts curriculum, as a math teacher you might reinforce past tense by pointing

out that when we talk about math operations or concepts we learned previously, we use the past tense of English, much like a history teacher might draw attention to past tense forms when discussing and reading about historical events. Similarly, English teachers teach morphology (base words, roots, prefixes, suffixes), but you may teach many words with these word parts as key vocabulary (such as *application* or *bivariate*). If English learners (and other students) have an opportunity to read, write, and orally produce these words during math and other subjects such as English-language arts, history, or science, the words are repeatedly reinforced. And, if this reinforcement occurs every school day, one can assume that English learners' mastery of English will be accelerated, as happens with repeated practice in any new learning situation. For a usable and informative list of English word roots that provide the clue to more than 100,000 English words, refer to pages 60–61 of *Making Content Comprehensible for English Learners: The SIOP® Model* (Echevarria, Vogt, & Short, 2008). This is a must-have list for both elementary and secondary teachers in ALL curricular areas.

Picture a stool with three legs. If one of the legs is broken, the stool will not be able to fulfill its function, which is to hold a person who sits on it. From our experience, an English learner must have instruction in and practice with all three "legs" of academic vocabulary (key vocabulary, process/function words, and words/word parts that teach English structure) if they're going to develop the academic language they need to be successful students.

How Is Academic Language Manifested in Classroom Discourse?

Our teachers come to class,
And they talk and they talk,
Til their faces are like peaches,
We don't;
We just sit like cornstalks.
(Cazden, 1976, p. 74)

These poignant words come from a Navajo child who describes a classroom as she sees it. Teachers like to talk. Just observe any classroom and you'll find that the teacher does the vast majority of the talking. That might be expected because the teacher, after all, is the most expert math person in the classroom. However, for students to develop proficiency in language, interpret what they read, express themselves orally and in writing, participate during whole-class and small-group instruction, and explain and defend their answers, they need opportunities to learn and use academic language. To promote more student engagement in classroom discourse, the Interaction component is included in SIOP® Model. The features of the Interaction component, which should be familiar to you at this point, include:

- Frequent opportunities for interaction with and discussion between teachers and students and among students, which encourage elaborated responses about lesson concepts
- Grouping configurations support language and content objectives of the lesson
- Sufficient wait time for student responses consistently provided
- Ample opportunities for students to clarify key concepts in L1 (native language) as needed.

These features promote balanced turn-taking both between teachers and students and among students, providing multiple opportunities for students to use academic English. Notice how each feature of Interaction encourages student talk. This is in considerable contrast to the discourse patterns typically found in both elementary and secondary classrooms. Most instructional patterns involve the teacher asking a question, a student responding, the teacher evaluating the response (IRE: Initiation-Response-Evaluation), or providing feedback (IRF: Initiation-Response-Feedback), followed by another teacher-asked question (Cazden, 1986; 2001; Mehan, 1979; Watson & Young, 1986). A typical interaction between a teacher and her students during a math lesson is illustrated in the following example:

T: Who can tell me what we call the longest side of a triangle?

S: A hypotimus.

T: Well, that's close. Who can help Rodolfo with, who can tell us?

S: Hypotenuse.

T: That's right. It's called the hypotenuse. Very good.

And so it goes, often for a good portion of the lesson. Notice that the teacher asked questions that had one correct answer with no reasoning or higher level thinking required, the teacher controlled the interchange, and the teacher evaluated student responses. Also note that the person who used the most academic language (*hypotenuse, triangle*) was the teacher. The students didn't need to use more than one or two words in response to the teacher's questions in order to participate appropriately. Only two students were involved, while the others sat quietly.

The IRE/IRF pattern is quite typical and it has been found to be one of the least effective interactional patterns for the classroom (Cazden, 1986; 2001; Mehan, 1979; Watson & Young, 1986). More similar to an interrogation than to a discussion, this type of teacher–student interaction stifles academic language development and does not encourage higher level thinking because most of the questions have a straightforward known answer. Further, we have observed from kindergarten through high school that most students become conditioned to wait for someone else to answer. Often it is the teacher who ultimately answers his or her own question, if no students volunteer.

In classrooms where the IRF pattern dominates (Initiation-Response-Feedback), the teacher's feedback may inhibit learning when she changes students' responses by adding to or deleting from their statements or by completely changing a student's intent and meaning. Because the teacher is searching for a preconceived answer and often "fishes" until it is found, the cognitive work of the lesson is often carried out by the teacher rather than the students. In these classrooms, students are seldom given the opportunity to elaborate on their answers; rather, the teacher does the analyzing, synthesizing, generalizing, and elaborating.

Changing ineffective classroom discourse patterns by creating authentic opportunities for students to develop academic language is critically important because as one acquires language, new concepts are also developed. Think about the previous example of visiting a financial planner. Each new vocabulary word you learned and understood (e.g., deferred annuity, fixed interest securities) is attached to a concept that in turn expands your ability to think about how to protect your hard-earned money. As your own system of word-meaning grows in complexity, you are more capable of understanding the associated concepts and generating the self-directed speech of verbal thinking: "When that Certificate of Deposit matures, I think I'll put it in a mutual fund." Without an understanding of the words and the concepts they represent, you would be incapable of thinking about (self-directed speech) or discussing (talking with another) financial planning.

Why Do English Learners Have Difficulty with Academic Language?

Developing academic language has proven to be quite challenging for English learners. In fact, in a study that followed EL students' academic progress in U.S. schools, researchers found that the ELs actually regressed over time (Suarez-Orozco, Suarez-Orozco & Todorova, 2008). There are myriad influences that affect overall student learning, and academic language learning in particular. Some factors, such as poverty and transiency, are outside of the school's sphere of influence, but let's focus on some of the influences that are in our control, namely what happens instructionally for these students that facilitates or impedes their learning.

Many classrooms are devoid of the kinds of supports that assist students in their quest to learn new material in a new language. Since proficiency in English is the best predictor of academic success, it seems reasonable that teachers of English learners should spend a significant amount of time teaching the vocabulary required to understand the lesson's topic. However, in a study that observed 23 ethnically diverse classrooms, researchers found that in the core academic subject areas only 1.4% of instructional time was spent developing vocabulary knowledge (Scott, Jamison-Noel, and Asselin, 2003).

The lack of opportunity to develop oral language skills hinders students' progress in all subject areas. Passive learning—sitting quietly while listening to a teacher talk—does not encourage engagement. In order to acquire academic language, students need lessons that are meaningful and engaging and that provide ample opportunity to practice using language orally. Successful group work requires intentional planning and giving students instructions about how to work with others effectively; teacher expectations need to be made clear. Grouping students in teams for discussion, using partners for specific tasks, and other planned configurations increase student engagement and oral language development.

Another related influence on language development is access to the language and the subject matter. Think about a situation in which you hear another language spoken. It could be the salon where you get a manicure or your favorite fast food place. Just because you regularly hear another language, are you learning it? Typically, not. Likewise, many English learners sit in class and hear what amounts to "English noise." It doesn't make sense to them and thus, they are not learning either academic language or the content being taught. Without the kinds of practices that are promoted by the SIOP® Model, much of what happens during the school day is lost on English learners.

Finally, some teachers have low expectations for EL students (Lee, 2005; NCTM, 2008). They are not motivated to get to know the students, their cultures, or their families. Poor performance is not only accepted, but expected. Rather than adjusting instruction so that it is meaningful to these students, teachers attribute lack of achievement to students' cultural background, limited English proficiency and, sadly, ability.

How Can We Effectively Teach Academic Language in Mathematics?

In a recent synthesis of existing research on teaching English language and literacy to ELs in the elementary grades, the authors make five recommendations, one of which is to "Ensure that the development of formal or academic English is a key instructional goal for English learners, beginning in the primary grades" (Gersten et al., 2007, pp. 26–27). Although few empirical studies have been conducted on the effects of academic language instruction, the central theme of the panel of researchers conducting the synthesis was the importance of intensive, interactive language practice that focuses on developing academic language. This recommendation was made based upon considerable expert opinion, with the caveat that additional research is still needed.

Because you are already familiar with the SIOP® Model, you know that effective instruction for English learners includes focused attention on and systematic implementation of the SIOP® Model's eight components and 30 features. Therefore, use the SIOP® protocol to guide lesson design when selecting activities and approaches for teaching academic language in mathematics.

Jeff Zwiers (2008, p. 41) notes that "academic language doesn't grow on trees." Rather, explicit instruction through a variety of approaches and activities provides English learners with multiple chances to learn, practice, and apply academic language (Stahl & Nagy, 2006). Teachers must provide comprehensible input (Krashen, 1985) as well as structured opportunities for students to produce academic language in their content classes. This will enable English learners to negotiate meaning through confirming and disconfirming their understanding while they work and interact with others.

In addition to explicit vocabulary instruction, we need to provide a variety of scaffolds, including context. Writing a list of math terms on the board or pointing out sentences that are bolded in the textbook only helps if students know what they mean. To create a context for learning academic English, teachers must preteach terms and sentence patterns (e.g., interrogative and declarative), and explain them in ways that students can understand and relate to, followed by showing how the terms and sentence patterns are used in the textbook. Scaffolding involves providing enough support to students so that they are gradually able to be successful independently. Another way of scaffolding academic English is to have word walls or posters displayed that show commonly used terms, operations, and math processes. Certainly, older learners can work in groups to create these posters with mnemonics, including cartoons or other illustrations. These aids reduce the cognitive load for English learners so that they can focus on math operations or processes without having to remember their associated linguistic terms. As students refer to and use these posted academic language words and phrases, they will internalize the terms and begin to use them independently.

In the lesson plans and units that appear in Chapters 3–8, you will see a variety of instructional techniques and activities for teaching, practicing, and using academic language in the mathematics classroom. As you read the lesson plans, reflect on why particular activities were selected for the respective content and language objectives. Additional resources for selecting effective activities that develop academic language and content knowledge include: Buehl's *Classroom Strategies for Interactive Learning* (2009); Vogt & Echevarria's *99 Ideas and Activities for Teaching English Learners with the SIOP® Model* (2008); and Reiss's *102 Content Strategies for English Language Learners* (2008).

Secondary teachers will also find the following books to be helpful: Jeff Zwiers's *Building Academic Language: Essential Practices for Content Classrooms (Grades 5–12)* (2008), and his book, *Developing Academic Thinking Skills in Grades 6–12: A Handbook of Multiple Intelligence Activities* (2004).

The Role of Discussion and Conversation in Developing Academic Language

Researchers who have investigated the relationship between language and learning suggest that there should be more balance in student talk and teacher talk in order to promote meaningful language learning opportunities for English learners (Cazden, 2001; Echevarria, 1995; Saunders & Goldenberg, 1992; Tharp & Gallimore, 1988; Walqui, 2006). In order to achieve a better balance, teachers need to carefully analyze their own classroom interaction patterns, the way they formulate questions, how they provide students with feedback, and the opportunities they provide for students to engage in meaningful talk.

Not surprisingly, teacher questioning usually drives the type and quality of classroom discussions. The IRE or IRF pattern discussed previously is characterized by questions to which the teacher already knows the answer and results in the teacher unintentionally expecting students to "guess what I'm thinking" (Echevarria & Silver, 1995). In fact, researchers have found that explicit, "right there" questions are used about 50% of the time in classrooms (Zwiers, 2008), and math lends itself to brief, factual exchanges.

In contrast, open-ended questions that do not have quick "right" or "wrong" answers promote greater levels of thinking and expression and are supported by the National Council of Teachers of Mathematics (NCTM). The teaching of mathematics has changed:

> (I)n many mathematics classrooms today, the emphasis has shifted from students acquiring vocabulary and solving standard word problems to learners explaining solution processes, describing ideas, presenting arguments, and proving conclusions. The teacher–student interactions imply a complex association between language and mathematics in general, and more specifically, the intricate nature of the relationship between learning mathematics and acquiring the language . . . in which mathematics is learned (Cuevas, 2005, p. 71).

During math lessons, there should be more of an emphasis on promoting classroom discourse by students questioning one another, reasoning rather than merely memorizing

FIGURE 1.1 *National Council of Teachers of Mathematics Communication Standards*

Instructional programs from prekindergarten through grade 12 should enable all students to:

- organize and consolidate their mathematical thinking through communication.
- communicate their mathematical thinking coherently and clearly to peers, teachers, and others.
- analyze and evaluate the mathematical thinking and strategies of others.
- use the language of mathematics to express mathematical ideas precisely.

procedures, making connections, solving problems and communicating solutions. In this way the NCTM communication standards, shown in Figure 1.1, can be met. For example, questions such as, "Would you add all these numbers or use multiplication? Why?" and requests such as, "Explain why we use arrows at the ends of a line. . . . Okay, what do they represent?" not only get students thinking about math processes, but also provide an opportunity for students to grapple with ideas and express themselves using academic English.

Something as simple as having students turn to a partner and answer a question first, before reporting out to the whole class, is an effective conversational technique, especially when the teacher circulates to monitor student responses. Speaking to a peer may be less threatening, and this method also gets every student actively involved. Also, rather than responding to student answers with "Very good!" teachers who value conversation and discussion encourage elaborated responses with comments like, "Can you tell us more about that?" or "What made you think of that?" or "Did anyone else have that idea?" or "Please explain how you figured that out."

Jeff Zwiers (2008, pp. 62–63) has classified the types of comments teachers can make to enrich classroom talk. By using comments like those that follow, you will be able to achieve a greater balance between student talk and teacher talk. Further, classroom interactions will be less likely to result in an IRE or IRF pattern. Try using some of these comments and see what happens to the interaction pattern in your own classroom!

To Prompt More Thinking

- You are on to something important. Keep going.
- You are on the right track. Tell us more.
- There is no right answer, so what would be your best answer?
- What did you notice about . . .

To Fortify or Justify a Response

- That's a good probable answer . . . How did you get to that answer?
- Why is what you said so important?
- What is your opinion (impression) of . . . Why?

To See Other Points of View

- That's a great start. Keep thinking and I'll get back to you.
- If you were in that person's shoes, what would you have done?
- Would you have done (or said) it like that? Why or why not?

To Consider Consequences

- Should she have . . .?
- What if he had not done that?
- Some people think that . . . is [wrong, right, and so on]. What do you think? Why?
- How can we apply this to real life?

A conversational approach is particularly well suited to English learners who, after only a few years in school, frequently find themselves significantly behind their peers in most academic areas, usually because of low reading levels in English and underdeveloped language skills. Students benefit from a conversational approach in many ways because conversation provides:

- A context for learning in which language is expressed naturally through meaningful discussion
- Practice using oral language, which is a foundation for literacy skill development
- A means for students to express their thinking, and to clarify and fine-tune their ideas
- Time to process information and hear what others are thinking about
- An opportunity for teachers to model academic language, use content vocabulary appropriately, and, through think-alouds, model thinking processes
- Opportunities for students to participate as equal contributors to the discussion, which provides them with repetition of both linguistic terms and thinking processes and results in their eventual acquisition and internalization for future use

A rich discussion, or conversational approach, has advantages for teachers as well, including the following:

- Through discussion, a teacher can more naturally activate students' background knowledge and assess their prior learning.
- When working in small groups with each student participating in a discussion, teachers are better able to gauge student understanding of the lesson's concepts, tasks, and terminology, as well as discern areas of weakness.
- When teachers and students interact together, a supportive environment is fostered, which builds teacher-student rapport.

When contemplating the advantages of a more conversational approach to teaching, think about your own learning. In nearly all cases, it probably takes multiple exposures to new terms, concepts, and information before they become yours to use independently. If you talk with others about the concepts and information you are learning, you're more likely to remember them.

English learners require even more repetition and redundancy. As they have repeated opportunities to improve their oral language proficiency, ELs are more likely to use English, and more frequent use results in increased proficiency (Saunders & Goldenberg, 2009). With improved proficiency, ELs are more adept at participating in class discussions. Discussion and interaction push learners to think quickly, respond, construct sentences, put their thoughts into words, and ask for clarification through classroom dialogue. Discussion also allows students to see how other people think and use language to describe their thinking (Zwiers, 2008).

Productive discussion can take place in whole class settings, but it is more likely that small groups will facilitate the kind of high-quality interaction that benefits English learners. Working to express ideas and answers to questions in a new language can be intimidating for students of all ages. Small group work allows them to try out their ideas in a low-stress setting and to gauge how similar their ideas are to those of their peers. Working with partners, triads, or in a small group also provides a chance to process and articulate new information with less pressure than a whole class setting may create.

Sharing conversational control with students involves some risk-taking on the part of the teacher as well as practice on the part of students who may prefer to answer questions with monosyllabic words. Simply telling students to "discuss" will likely have poor results. We need to teach students how to engage in meaningful conversation and discussion and provide the structure and support they need to be successful. Rather than sitting as "quiet cornstalks," students, including English learners, can learn to express themselves, support their answers, advocate their positions, and defend their beliefs. When this occurs, we establish a classroom environment in which conversational control is shared among teachers and students alike.

What Is the Academic Language of Mathematics?

There are myriad terms that are used in academic settings. As mentioned previously, some of these are used commonly across content areas and others are content specific. The metaphor of bricks and mortar is helpful here as we think of some words representing bricks, such as math content-specific words (*algebraic symbol, formula, and geometric shape*). The mortar refers to general academic words such as *describe, represent, and approximate* (Dutro & Moran, 2003). Understanding both types of terms is often the key to accessing content for English learners. For example, although most students need to have terms related to math explicitly taught, English learners also require that general academic words be included in vocabulary instruction.

As you plan for lessons that teach and provide practice in both math-specific academic language and more general academic language, use your teacher's guides from the math text to note the highlighted vocabulary, as well as other terms and phrases that may need to be taught. Also, you may use English language arts content standards and your state English language development standards for ELs to assist you in selecting the general academic language you need to teach and reinforce. Other resources include the "1,000 Most Frequent Words in Middle-Grades and High School Texts" and "Word Zones™ for 5586 Most Frequent Words," which were collected by Hiebert (2005) and may be found online at www.textproject.org. For those of you who are high school teachers, you might also want to take a look at the Coxhead Academic Word List (Coxhead, 2000).

In the study of mathematics, students are exposed to many new terms that they most likely won't encounter anywhere other than in math class. These terms are found in the textbook and ancillary materials, presented during lessons, and found in math standards. Let's take a look at the various terms that are present in a few selected NCTM math standards. The words that are math-specific are **bolded** and general academic words are underlined:

In Pre-Kindergarten through Grade 2, All Students Should:

- understand <u>situations</u> that entail **multiplication** and **division,** such as **equal** <u>groupings</u> of <u>objects</u> and <u>sharing equally</u>.
- <u>model</u> situations that involve the **addition** and **subtraction** of **whole numbers,** using <u>objects</u>, <u>pictures</u>, and <u>symbols</u>.

In Grades 3–5, All Students Should:

- <u>identify</u> such **properties** as **commutativity, associativity,** and **distributivity** and use them to **compute** with **whole numbers.**
- <u>propose</u> and <u>justify conclusions</u> and <u>predictions</u> that are based on **data** and <u>design studies</u> to further <u>investigate</u> the <u>conclusions</u> or <u>predictions</u>.

In Grades 6–8, All Students Should Be Able to:

- <u>compare</u> and <u>order</u> **fractions, decimals,** and **percents** <u>efficiently</u> and find their <u>approximate</u> locations on a **number line.**
- understand both **metric** and **customary systems of measurement**.

In Grades 9–12, All Students Should:

- use **number-theory** <u>arguments</u> to <u>justify</u> relationships involving **whole numbers.**
- understand the meaning of **equivalent forms of expressions, equations,** <u>inequalities</u>, and <u>relations</u>.

As you can see, many of the underlined words may be used in other content areas as well, but students need to be explicitly taught their meaning. Some of these words are common, but have a specialized meaning in math. For students who speak a Latin-based language such as Spanish, cognates may help in teaching some words. For example, *predict* in English is *predecir* in Spanish; *justify* in English is *justificar* in Spanish; *multiply* in English is *multiplicar* in Spanish. Math-specific words should be explicitly taught as part of each math lesson.

In Appendix B you will find a comprehensive list of academic math vocabulary words and phrases found in the NCTM content and process standards organized by the grade-level clusters used throughout this book (K–2, 3–5, 6–8, 9–12). Your state's standards and domains may differ a bit, but we hope this extensive list will assist you in your lesson and unit planning, and in the writing of your content and language objectives.

Concluding Thoughts

Proficiency in English is the best predictor of academic success, and academic language proficiency is an important part of overall English proficiency. In this chapter we have discussed what academic language is, why it is important, and how it can be developed in math classes and across the curriculum. In all content areas, teachers need to plan to explicitly teach both content area terms and general academic terms as well as provide opportunities for students to develop other aspects of academic language. In the mathematics classroom, we do so in order that English learners can fully participate in lessons, acquire concepts and processes of math, and increase their academic language proficiency.

An important way to provide opportunities for students to learn and practice academic language is through classroom conversations and structured discussions. When you teach students how to participate in classroom conversations, you not only improve their English skills but also prepare them to understand the type of language used by historians, scientists, mathematicians, authors, literary critics, and other scholars. You will give them the tools they need to have practice with language skills that enable them to back up claims with evidence, be more detailed in their observations, use persuasive language compellingly in arguments, and compare points of view, with the result being academic achievement and school success.

chapter 2

SIOP® Lesson Planning and Unit Design

The process of careful, thoughtful lesson and unit planning is one of the most important factors that contribute to effective teaching. We frequently hear from SIOP® teachers that they have come to appreciate the importance of the planning process for teaching lessons that engage their students and result in improved student performance. In this chapter, we provide specific information for developing effective SIOP® lesson plans for mathematics, and the decision-making process teachers go through when designing effective SIOP® mathematics units. We also discuss the advantages of collaboration.

We hope that the information in this chapter will assist you in understanding the lesson and unit planning process more deeply and will provide you with ideas for your own planning and teaching.

Lesson Planning

SIOP® Lesson Plan Formats

As a teacher becomes more and more familiar with SIOP® lesson planning—incorporating the features of the SIOP® Model into every lesson—there is usually a particular lesson plan format that is favored. At present, there are numerous SIOP® lesson plan formats that teachers are using in schools and districts throughout the country. Some have been created by teachers, others have been adapted for the SIOP® Model from district lesson plans, and a number have been created by SIOP® *National Faculty* and the SIOP® authors. As you know, four lesson plan formats are included in the core text, *Making Content Comprehensible for English Learners: The SIOP® Model* (Echevarria, Vogt, & Short, 2008; 2010a; 2010b), and eleven other lesson plan formats are included in the book *Implementing the SIOP® Model through Effective Coaching and Professional Development* (Echevarria, Short, & Vogt, 2008). You will find four of the lesson plans in writable electronic lesson plan formats at *www.siopinstitute.net*. While we do not endorse any particular lesson plan format, we advocate that you select one that works well for you.

The SIOP® lesson plan format that is used throughout this book for elementary lessons was adapted from one created by Melissa Castillo and Nicole Teyechea. It has a Key along the top to identify the acronyms used and it is a bit more conceptual in orientation. The SIOP® lesson plan used for secondary lessons has been adapted by Araceli Avila from the version on page 230 of *Making Content Comprehensible for English Learners: The SIOP® Model* (Echevarria, Vogt, & Short, 2008). It has an area to check off the SIOP® features included in the lessons and has a step-by-step description of the lesson. We have found that teachers tend to adjust lesson plan formatting to suit their own teaching style.

You will find that each section of the two lesson plan templates provides guidance on things to consider or questions to ponder as you write lesson plans.

Take note of the kinds of questions to consider in planning that address both content learning and language/literacy development. For example, since oral language practice is an important part of instruction for ELs, you should consider what kinds of frames to use so that students participate using complete sentences. Another important aspect of effective teaching is assessing student learning. The template prompts you to decide on the formal and informal assessment that will be used with each activity. Also, writing a step-by-step description of what you will do and what the students will do to meet the lesson's learning objectives helps to make sure that the lesson will flow smoothly. For example, as you write a step in the lesson, you may realize that successful completion of that step would be enhanced by some type of scaffold for students such as a visual display of the information or a worksheet that shows a completed sample problem. Or, you may realize that an activity requires materials that you will need to have on hand.

SIOP® LESSON PLAN *(Elementary Grades)*

Key: SW = Students will; TW = Teacher will; HOTS = Higher Order Thinking Skills (questions and tasks)

Lesson: *Grade level, Unit, and Title*
Grade:

NCTM Standard: *Note the standard listed is from the National Council of Teachers of Mathematics. List the appropriate standard for your state and grade level.*
Expectation: *List the performance objective, indicator, or student outcome that supports the state standard.*

Visuals & Resources: *What additional resources do you need? Include those that are outside of the "regular" curriculum.*

Key Vocabulary:	General Frames:
List key words students must know in order to understand the lesson/concept. Include words they must master for ongoing learning.	*How are you going to ensure that students are using complete sentences? Provide language structures that can be used for multiple lessons. Provide students with functional language practice.*
HOTS: *Consider: What higher order thinking questions will you ask? What higher order thinking tasks will students participate in?*	*For example: I agree/disagree with . . . I think . . . I understand . . . I will . . . I can_____ because . . .*
	Specific Frames: *How will you ensure that students have the language to respond to questions related to the specific content being taught? Provide language structures for students to use in one lesson. What language do they need to answer "HOTS" questions? Frames should contain content-specific vocabulary.*

Connections to Prior Knowledge/Provide Background Information: *How are you going to connect to students' own experiences and prior learning? What questions might you ask? What activity are you going to use to involve students and build connections with the new concepts? How are you going to introduce or review the key vocabulary?*

Content Objectives:	Meaningful Activities:	Review/Assessment:
What they need to have learned at the end of the lesson. Should be aligned to a state standard/outcome or indicator. Must be measurable.	1. *List the activities you will do as a teacher and what students will be doing: I do, We do, and You do.*	1. *List both formal and informal assessments that will be used for each activity.*
Language Objectives:	2. *Number or bullet activities and be sure to align them with the appropriate content or language objectives and assessments.*	2. *List Teacher's Manual pg #s when the assessment is from the book or workbook.*
Consider: How will the students practice/apply key concepts, and academic language using reading, writing, speaking, and/or listening skills? How are you ensuring that language or English language development (ELD) standards are being taught and practiced? Objectives also have to be measurable.	3. *Consider the following as you plan your meaningful activities: Are students using learner strategies? Are they being challenged? Are they using language? Are they interacting? Are they engaged?* *List Teacher's Manual pg #s when the activity is from the book.*	3. *Whatever students are doing should provide you with information on whether student learning and understanding is taking place.*

Wrap-up: *Always review objectives! What activity are you going to use to close and review key concepts or vocabulary? (e.g., outcome sentences, journal, ticket out).*

SIOP® LESSON PLAN *(Secondary Grades) Lesson's Title*

Class/Subject Area(s): *Subject*   **Grade Level:** *Elementary, Middle School, or High School*
Unit/Theme: *Unit's Big Idea*   **Lesson Duration:** *Number of Class Periods or Total Minutes*

NCTM Standards

Content	*Process*
List of appropriate grade-level content based on state standards. *Note: The content and process standard(s) listed and checked off in the lesson plans are from the National Council of Teachers of Mathematics.*	☐ Problem Solving ☐ Reasoning & Proof ☐ Communication ☐ Connections ☐ Representations

Content Objective(s):
Statement(s) written in student-friendly language that makes reference to what students will learn by the end of the lesson. Content objective(s) must be aligned to state standards and must be measurable.

Language Objective(s):
Statement(s) written in student-friendly language, which ensures students will practice/apply key concepts and utilize academic language by using the language skills of reading, writing, speaking, and/or listening. Language objective(s) supports the content objective(s) and must be measurable.

Key Vocabulary:	*Supplementary Materials:*
Content Vocabulary **Functional Vocabulary** *List of content and functional key words students must know in order to understand the lesson/concept. Words they must master for ongoing learning.*	*List of additional resources, which include those that are outside of the "regular" curriculum.*

SIOP® Features: *(Teacher checks off features incorporated in the lesson sequence.)*

Preparation	**Scaffolding**	**Grouping Options**
__ Adaptation of content	__ Modeling	__ Whole class
__ Links to background	__ Guided practice	__ Small groups
__ Links to past learning	__ Independent practice	__ Partners
__ Strategies incorporated	__ Comprehensible input	__ Independent

Integration of Processes	**Application**	**Assessment**
__ Reading	__ Hands-on	__ Individual
__ Writing	__ Meaningful	__ Group
__ Speaking	__ Linked to objectives	__ Written
__ Listening	__ Promotes engagement	__ Oral

Lesson Sequence:

Step-by-step description of what teacher and students will do to meet learning objectives.

Reflections:
After teaching the lesson, the teacher reflects on what worked, what did not work, and what revisions, additions, and/or deletions need to be made.

Taking time to write a lesson plan using the prompts on the templates will actually save time that may be lost in trying to locate materials, in having to explain something multiple times that showing a visual would have taken care of, or in reteaching later because student comprehension wasn't assessed as the lesson unfolded.

Our math contributors selected the lesson template format they were most comfortable using and we encourage you to do the same. Once you know the format and have practiced with it, lesson planning will be much easier. The most important consideration in selecting a lesson plan template is that it include all eight components and thirty features of the SIOP® Model. As you can see, although the elementary and secondary templates differ in format, they both include the following elements:

- Unit title, SIOP® lesson topic, and grade level
- Relevant content standards
- Key vocabulary
- Supplementary materials
- Content objectives
- Language objectives
- Meaningful activities: Lesson sequence
- Plans for review and assessment throughout the lesson
- Wrap-up

The acronyms used throughout are:

- SW = Students will . . . Other similar terms used by teachers are SWBAT (Students will be able to . . .) and TSW (The student will . . .)
- TW = Teacher will . . .This is used to describe what the teacher does during the lesson.
- CO = Content Objective (See Appendix A for a brief discussion of objectives.)
- LO = Language Objective (See Appendix A for a brief discussion of objectives.)
- HOTS = Higher order thinking skills.

In Chapters 3 and 4, we use these lesson plan templates to show how meaningful activities can be integrated into math lessons, and in Chapters 5 through 8, we present comprehensive one-week units using the templates.

As you may have discovered already, implementing the features of the SIOP® Model to a high degree will initially require careful, detailed planning. However, as shown in Figure 2.1, as you practice including the features in lessons, eventually you will find that the SIOP® Model becomes a way of teaching and less detailed lesson plans are needed. As such, we recommend that you consider SIOP® lesson planning as a process that develops over time.

What's in a SIOP® Lesson?

Effective teaching includes explicit instruction of the content material and academic vocabulary, teacher modeling, guided practice, practice and application, and independent practice. In this process, seen in Figure 2.2, which is also referred to as gradual release of responsibility, the teacher presents information and models "the type of thinking required to solve problems, understand directions, comprehend a text, or the like" (Fisher & Frey, 2008, p. 5).

A teacher-led focus lesson including ample modeling is followed by guided instruction, in which the teacher carefully monitors student application and practice of the

FIGURE 2.1 *SIOP® Lesson Planning Over Time*

Detailed SIOP® Lesson Planning

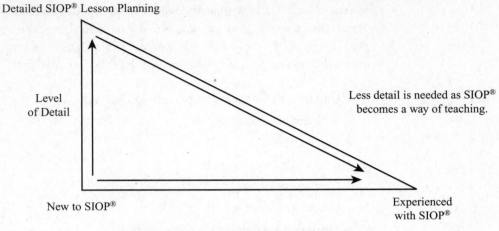

Level of Detail

Less detail is needed as SIOP® becomes a way of teaching.

New to SIOP®

Experienced with SIOP®

Experience Using the SIOP® Model

FIGURE 2.2 *Gradual Release of Responsibility*

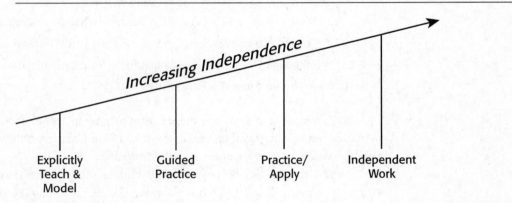

Increasing Independence

Explicitly
Teach &
Model

Guided
Practice

Practice/
Apply

Independent
Work

From: Echevarria, J., Vogt, M.E. & Short, D. (2004). *Making Content Comprehensible for English Language Learners: The SIOP Model, Second Edition.* Boston: Allyn & Bacon.

information presented. Then students are given an opportunity to collaborate. During this phase, students discuss ideas and information they learned during the focus lessons and guided instruction—not new information. Finally, when students have acquired sufficient knowledge from the first three phases and are ready to use the information with independent tasks, they are given the opportunity to build understanding of mathematical concepts through practice and application of those concepts. For English learners, this process also includes the development of academic English through language and literacy practice opportunities.

As you will see in the lessons, this is not a lock-step process, but a way for teachers to purposefully shift responsibility for learning to students and may be done in a variety of ways. Perhaps most importantly, each lesson reflects all eight of the SIOP® components, including the language and content objectives that drive instruction.

From our own experience as SIOP® teachers and SIOP® professional developers, we know the importance of making content and language objectives a relevant part of lessons. The objectives help teachers stay focused on the point of the lesson and, if used meaningfully, they provide students with guidance for learning. In particular, language objectives are critically important to the development of academic language.

In Figure 2.3 we have suggested a number of activities for making content and language objectives a relevant part of daily lessons. We hope you find them useful in your own teaching practice.

In addition, you can review the six types of language objectives in the core SIOP® texts (Echevarria, Vogt, & Short, 2008, pp. 29–30; 2010a, pp. 32–33; 2010b, pp. 32–33). The content and language objectives included in the sample lesson plans in Chapters 3–8 will also provide ideas for teaching the math content and academic language English learners must practice in order to be able to meet the content standards.

FIGURE 2.3 *Activities for Meaningful Objectives*

Read the objective as a shared reading piece with your entire class. Then ask them to paraphrase the objective with a partner, each taking a turn.

Ask students to read the objectives on the board and to add them to their Math Learning Notebooks in a paraphrased form. Then have them read their paraphrased objectives to each other. This could be done as a sponge in the first part of the class while you are taking roll!

Present the objective and then do a **Timed Pair Share,** asking students to predict some of the things they think they will be doing in class that day.

Ask students to do a **Rally Robin** (taking turns), naming things they will be asked to do that day in that particular class.

Ask students to pick out important words from the objective, such as the action words or nouns, and highlight them.

Give students important words to "watch" and "listen" for during the lesson and call their attention to that part of the objective when you mention the academic vocabulary in the lesson.

Reread the objective using shared reading during the lesson to refocus your students.

Ask students to write one or two sentences explaining what they learned in class today and to show an example. This can be done in a learning log or on a sticky note left on the desk.

Round Robin (taking turns talking for a specified time with a partner) about how students can prove they met their learning objective for the day.

Ticket-Out—students write a note to the teacher (or a letter to the parent at the end of each day) telling her/him what they learned and asking any clarifying questions.

Source: Castillo, 2008.

Using the SIOP® Protocol for Planning Purposes

After learning a new approach to teaching, the natural tendency is to fall back into familiar patterns of practice and, as time passes, abandon the newly learned approach. Using the SIOP® protocol regularly as a planning tool helps to keep the features of the SIOP® Model familiar. The more you read through and think about the thirty features, the more comfortable you will become in naturally including them in lessons. Over time, incorporating the features in lessons becomes a habit.

The protocol can also be used for self-evaluation; it provides concrete features that keep teachers accountable for implementing the SIOP® Model with fidelity. A practice at one school that implemented the SIOP® Model with impressive results, including improved student achievement, was to place copies of the SIOP® protocol in the mail room for teachers to take. This kept the SIOP® protocol "on their radar" regularly. The SIOP® coach at the school commented that the accountability the protocol provided encouraged teachers to continually improve their practice. She said that teachers didn't just go back into their classrooms, close the door, and do whatever they wanted. Instead, they were motivated to change their teaching to include practices that are best for English learners—those practices reflected in the features of the SIOP® protocol (Echevarria, Short, & Vogt, 2008).

We encourage you to use the SIOP® protocol as a checklist during lesson planning. As we mentioned, the more familiar you become with the instructional features on the SIOP® protocol, the more you will deepen your understanding of effective practices for English learners and transform your teaching.

Technology and SIOP® Lessons

The increasing use of technology in the classroom is an exciting trend, filled with many possibilities. A growing number of teachers use interactive white boards (I-boards) as replacements for traditional whiteboards, overhead transparencies, or flipcharts or video/media systems such as a DVD player and TV combination. A number of interactive software programs such as Inspiration, Kidspiration, and Language Learner are available, as are interactive websites. Some interactive whiteboards allow teachers to record their instruction and post the material for review by students at a later time, providing much-needed opportunities for repetition for English learners. Other technologies include hand-held devices that can be used in myriad ways such as recording student responses, learning about concepts in measurement, practicing multiplication tables, and taking notes in class, just to name a few.

However, not all schools have these resources available for teachers because of lack of funds or a low commitment to technology use. Often, it is the schools with high numbers of English learners that are the most disadvantaged. In order to be sensitive to teachers of ELs who lack access to more advanced technologies, the lesson plans presented throughout this book are fairly "low tech." If you are one of the fortunate teachers who have these resources at your disposal, when the lesson plan calls for use of an overhead transparency or worksheet, you would use an interactive white board; when the lesson plan suggests showing a picture, you might project a website. We invite you to adapt the lessons to the level of technology available to you.

Unit Design

Joanne Monroe is a teacher who is contemplating the design of her next SIOP® unit plan. Like every teacher, her planning is guided by a number of considerations. First, there is a math curriculum sequence she must follow to meet the district's scope and sequence. This sequence is supported by a district-adopted textbook and other assigned materials. Also, the district's pacing charts dictate how much time should be spent on each topic. Finally, Joanne must ensure that all state standards are addressed in her lessons and are met by her students.

In reviewing her math textbook and materials, Joanne realizes that they do not clearly align with the standards, nor do they highlight and emphasize enough the academic vocabulary that she knows is crucial for her EL students' academic success. In order for her students to meet the standards, she will have to fill in the gaps of the curriculum.

Another challenge Joanne faces is how to teach concepts in a way that makes them understandable for her students. For English learners who are studying new material in a new language, the textbook introduces concepts in a very abstract manner and doesn't provide students with the opportunity to practice new content knowledge concretely. Joanne is also concerned that the textbook moves students very quickly from theory to application without ample opportunity to develop a strong understanding of the concepts. From experience, she has learned that this is a recipe for failure, especially for English learners. Joanne's professional training and teaching experience have helped her understand the importance of providing both opportunities for English learners to develop academic and functional vocabulary and time for students to develop an understanding of mathematical concepts, rules, and processes. She also knows that ELs benefit from explicit teaching that uses concrete materials, modeling, guided practice, and independent practice for teaching concepts and academic language. These instructional practices that help English learners succeed aren't always reflected in district or state guidelines.

In short, Joanne's dilemma is how to bring together what she knows is best for her students and the curriculum she is expected to implement in a way that will ensure academic success for all her students. Joanne has been trained in the SIOP® Model and will use what she has learned to begin planning for effective instruction. She will use the components and features of the SIOP® Model as a framework for planning and teaching math lessons. In this way she can take what she is required to teach and turn it into a plan for effectively meeting the academic and language needs of all her students.

In order to get started with her Math units, Joanne has made a list of questions to guide herself in determining what the learning goals are and how all students will attain them. She asks herself the questions in Figure 2.4.

As you can see, the questions in Figure 2.4 assist a teacher in taking a broad view of the unit; she considers the best ways to move students to higher levels of understanding of the math concepts and academic language in the unit, including the materials needed, the language skills required, the background necessary to be successful, and ways to assess understanding.

In Chapters 5–8 you will find examples of math units that are the result of the planning process reflected in the preceding questions. As you read through each unit, note how we addressed our guiding questions by including meaningful activities that clearly embed each SIOP® component and feature in the lessons and how we determined the needs of our English learners and made modifications for them. Ensuring that lessons

FIGURE 2.4 *Guiding Questions for Unit Planning*

1. What must students learn in this unit?
2. How will I assess their learning of the unit's content concepts?
3. Where do I need to start to make sure students meet the learning goals identified in Question 1?
4. Is the time allotted sufficient to ensure that all my students meet the content and language objectives?
5. How many lessons will it take to teach the concepts and lead students to mastery?
6. What materials will I use to supplement the textbook?
7. What vocabulary must they already know and what additional vocabulary must be taught (content, process/ function, word parts from Chapter 1 p. 8)?
8. How will I make the lessons meaningful so students are engaged and motivated to learn?
9. How will the language skills of reading, writing, listening, and speaking be incorporated so all students have ample opportunity to practice not only the content knowledge but also the academic vocabulary of the unit?
10. What additional content concepts might have to be taught to students who have had their education interrupted or who might lack the English proficiency needed to comprehend the content and language objectives?

teach standards-based math concepts and skills while developing language and literacy can be challenging. Facility in this kind of lesson planning is often expedited when teachers collaborate in lesson and unit planning. Much of the planning for the units in these chapters was done collaboratively.

Collaborative Planning

So far, we have discussed individual factors in lesson planning: selecting the SIOP® lesson plan format you will use, choosing the questions you might consider when writing a lesson plan, and determining the kinds of questions you'll consider to guide your thinking about the learning goals you have for your students and how they will attain them. We now turn to a discussion of collaboration because SIOP® teachers have reported that their own ability to plan more effective lessons and implement the SIOP® Model in more sophisticated ways increased when they worked together with other teachers.

One of the benefits of collaborative planning is having the opportunity to receive input from colleagues who have reviewed your lesson plans. What we believe is perfectly clear to us—the instructions for an activity or the steps for solving a problem—isn't always clear to others. And if it isn't clear and understandable for teachers who know how lessons work, how can it be comprehensible for learners? The interaction, discussion, and questions that colleagues ask one another lead to lessons that are clear and precise.

Another advantage is that input from others increases variety in lessons. Teachers often tend to rely on their regular practices and tendencies (also known as being in a rut!). For example, a teacher may frequently use quick writes or tickets out as a way to review information with their students. In the process of collaborative planning, other ideas are generated and students benefit from the resulting variety in review techniques used. More importantly, there may be an instructional area that is neglected when a teacher has a preference for certain activities or assignments. Perhaps a teacher's tendency is to rely on teaching from the textbook and whole class discussion of math concepts followed by individual written work. In collaborating, other teachers may suggest activities that involve partner work or group work that requires oral practice using academic English. Further, when discussing teaching ideas with colleagues, SIOP® teachers have reported that "It also helps in that you learn how they teach certain content units, and how they transform their lessons into SIOP® lessons with a variety of activities" (Short, Vogt, & Echevarria, in press).

Further, collaboration helps teachers learn to fine tune their own lesson planning process. Sometimes trying to plan creative, interesting, cohesive math units week in and week out can be difficult. As one SIOP® teacher said, "Sometimes in planning units, I had reached a dead-end and thought I would have to completely start over. After talking with my partner, I would come away with fresh ideas and ways to improve what I was already working on" (Short, Vogt, & Echevarria, in press). The synergy created by co-planning benefits teachers and their students alike because of the refined lessons that are the result.

Sometimes when teachers get serious about improving their teaching and the quality of the lessons they offer their students, they take collaboration to the next level. Effective SIOP® teachers have told us that they would strongly encourage teachers to find a partner or small group of like-minded teachers at their school site to work with (Vogt, Echevarria & Short, in press). This type of collaboration is referred to as Professional Learning Communities (PLCs) or learning teams. A PLC is a group of teachers who work together (sometimes with administrators) to improve student learning through joint activities such as effective lesson planning and delivery and analysis of instruction, and who use student data to inform their decision making.

Of course, the challenge is in finding time for collaborative work. Some ways that SIOP® teachers have been able to carve out time for collaboration include: during grade-level meetings (elementary) or department meetings (secondary), when teachers have a similar prep time, during occasional sub release time, as part of planning time, or during curriculum mapping time. In optimal situations, structured time is provided for teachers to work together in PLCs. Focused time and energy needs to be applied when creating high-quality SIOP® lessons and units. But the most important aspect may be having the opportunity to come back together after the lessons are taught and to reflect on them while receiving feedback from a partner or group.

This advanced level of collaboration would include some of the following practices:

- Teacher self-reflection
- Peer observation
- Videotaped lesson analysis
- Group discussion of SIOP® lessons
- SIOP® coaching

During PLC meetings, teachers might discuss the strengths of a lesson (one that was observed, videotaped, or in lesson plan form), the ways that the lesson could be improved, and how the lesson might be altered to enhance student achievement. This process leads to teachers refining how they think about the features of the SIOP® Model itself because at first teachers tend to think about the SIOP® features at a superficial level and over time they increase their depth of understanding.

Not all professional growth activities are as extensive as those mentioned above. Greater understanding of instruction for ELs can be accomplished in other ways.

● . . . .

26

For example, at every staff meeting some time may be set aside for a discussion of some aspect of the SIOP® Model. For instance, the group may discuss language objectives one week and suggest a new strategy or skill to work on another week. The main purpose for the discussion is to keep the focus on teacher reflection and growth.

For a more in-depth discussion of effective collaboration and professional growth, please see *Implementing the SIOP® Model through Effective Professional Development and Coaching* (Echevarria, Short & Vogt, 2008).

Concluding Thoughts

In this chapter, we provided you with information for planning SIOP® math lessons and units by yourself and through collaboration with others. Although the emphasis of the chapter was on lesson planning and collaboration, the guiding questions ensure that the lessons incorporate both math teaching and language and literacy development.

As a mathematics teacher, you may think that planning lessons that also develop language and literacy skills is not part of the domain. However, language and literacy permeate all content areas. We know that the relationship between literacy proficiency and academic achievement grows stronger for students as they progress up the grade levels. In secondary school classes, language use becomes more complex and more content-area specific (Biancarosa & Snow, 2004), as you can see in the NCTM terms in Appendix B. To be successful in school, English learners are required to develop literacy skills for each content area <u>in</u> their second language as they simultaneously learn, comprehend, and apply content area concepts <u>through</u> their second language (Garcia & Godina, 2004).

Through the type of effective lesson and unit planning presented in this chapter—along with effective lesson delivery—we expect that these students will receive a strong foundation in academic English if they are enrolled in our elementary schools, and we want teachers to have the skills and knowledge base to continue that academic language and literacy development throughout the secondary school years.

We hope that through the guiding questions on the lesson plan templates, the unit planning discussion and the emphasis on collaborative planning, you will recognize the important role language and literacy play in mathematics, and that you will gain knowledge and ideas so you can design and deliver dynamic, engaging math lessons that promote content and language learning among your students.

Activities and Techniques for Planning SIOP® Mathematics Lessons

Lesson Preparation, Building Background, Comprehensible Input

Araceli Avila, Melissa Castillo and Jana Echevarría

Tomas Estorga is in his third year of teaching, and his principal will be conducting an observation of his class next week. His school has recently received in-depth training in the SIOP® Model and all teachers have begun implementing SIOP®-based lessons in their classrooms. The principal will be using the SIOP® protocol to document his observations and to provide Tomas with feedback. Obviously, Tomas wants to teach an effective SIOP® lesson while he is being observed. He is not struggling in the area of ideas, activities, or

techniques to choose from for his lesson; actually, he is overwhelmed by the wealth of resources he has accumulated from the teacher preparation classes, inservice sessions, and conferences he's attended. The biggest challenge for Tomas since he started teaching is how to pull the different techniques into a comprehensive lesson that will ensure students effectively meet the learning goals.

Introduction

Simply adding new techniques and activities to a lesson most likely won't give students the kind of cohesive, effective instruction they need to make progress academically. It's important that as teachers plan, they consider not only the content concepts that are the focus of their lessons but also the academic language that must be developed. In order for students to be successful, the activities they are expected to participate in must ensure that there is meaningful practice and application of both the content concepts and the lesson's academic language.

In the next two chapters we provide a variety of proven techniques and activities teachers can use in their math lessons so that students will be able to actively take part in the learning process, access the content, and develop proficiency in academic English. In the lesson plans provided, you will note that activities are embedded within a focused, thoughtful lesson plan. The lesson plans are used to provide students the opportunity to process content concepts as well as practice the language that makes up the content.

The activities themselves are not the lesson. As you read in Tomas's vignette, he, like many teachers, struggles with trying to figure out how to make the activities become a meaningful lesson. In actuality, the focus should be on the learning objectives that the activities are meant to support. Activities and techniques allow students to practice new learning and/or reinforce new concepts, information, or processes. They are important aspects of the lesson, but the language and content learning is the heart of the lesson.

The following list describes the information included in each technique.

Name of Activity or Technique In cases where we have drawn from the techniques in the *99 Ideas* book, we indicate that here. In other cases, our contributors have included techniques and activities that they have named or ones that are common to many ESL educators.

SIOP Component Here we identify the SIOP® component that this technique or activity addresses in the application we present. We recognize that several of these techniques may be used with different components as well.

Grade Level Often we provide a range of grade levels suitable for the technique or activity. Although the classroom application is grade-specific, teachers will likely be able to modify the technique to a different grade within the range suggested.

Grouping Configurations We explain what type of student configuration is most effective for this technique or activity. As you know, the way students are grouped is an important component of the SIOP® Model. We recommend deliberate, thoughtful groupings of students that match a lesson's objectives. Teachers may group students in pairs to promote more conversation and lower anxiety levels. They may create small groups of students with different abilities to give more advanced students a chance to act as the teacher and less proficient students an opportunity to have peer role models. At times, they may group students by first language so they can process the new information in a language they are more comfortable with before completing a task using English. There will also be times when a teacher wants to present new information to the whole group, but wants to ensure that it is comprehensible and the students stay focused.

Approximate Time Involved This information gives the teacher a sense of how long the activity may take in a lesson. It is useful for considering the pacing while preparing a lesson plan. It does not include the time needed to prepare for the activity.

Materials Materials needed for the activity are listed here.

Description/Procedure We describe the technique or activity here, including its general purpose and the steps of the procedure. Some suggestions for topics pertinent to specific subjects may be included, such as using a kinesthetic activity to teach geometric shapes and adding integers.

Application In this section, we explain how the technique or activity might be used in one specific class (with the grade band identified). The lesson concept is listed and content and language objectives that this technique can help the students meet are presented. As needed, key vocabulary terms are identified, particularly when the language objective is related to vocabulary learning. Then a classroom vignette illustrates how a teacher would use the technique or activity in the lesson.

Differentiation options Many of the techniques lend themselves to differentiation. As appropriate, we provide some suggestions for modifications. The adjustments might help apply the technique to students who are at different proficiency or grade levels, or to students who are under-schooled or who have wide gaps in their formal educational backgrounds.

The acronyms used throughout the chapters are:

- SW = Students will . . . Other similar terms used by teachers are SWBAT (Students will be able to . . .) and TSW (The student will . . .)
- TW = Teacher will . . . This is used to describe what the teacher does during the lesson.
- CO = Content Objective (see Appendix A for a brief discussion of objectives)
- LO = Language Objective (see Appendix A for a brief discussion of objectives)
- HOTS = Higher order thinking skills

Each of the techniques presented in these chapters may be used across various grade levels in K–12; they are not restricted to the grade level shown in the sample lessons. The detailed lesson plan is specific to a grade band, showing how the technique may be used to support the objectives of a particular lesson.

In our own practice of planning and delivering effective SIOP® instruction, we always begin with explicit content and language objectives that are derived from content-specific standards and the academic language needs of our English learners. We also consider what the final product, outcome, or academic expectation is for all students based on their literacy and language abilities. The sample lesson plans included in these chapters demonstrate how the objectives drive the selection of meaningful activities and vocabulary that must be developed, the process the teacher will take to guide students in mastering the concepts and language, and the way students will demonstrate their understanding and progress in meeting the learning goals.

As a way of organizing Chapters 3 and 4 we have arranged the techniques around SIOP® components (Chapter 3: Lesson Preparation, Building Background, and Comprehensive Input; Chapter 4: Strategies, Interactions, etc.). However, in order to create an effective SIOP® lesson plan, it is important to include all components and features. Note that in the plans, we have included techniques and activities for students to build background, apply learner strategies, answer higher order questions, use hands-on materials and manipulatives, interact with the teacher and other students, review and assess key content concepts and vocabulary, and so forth. We have also planned for the teacher to deliver instruction and facilitate student learning that supports both the content and language objectives, with a variety of techniques to make the content comprehensible, and multiple indicators to assess student comprehension.

The point we are trying to make emphatically is that all components of the SIOP® Model and its 30 features work together to make a lesson effective for English learners. Although we have not identified and isolated each and every feature in the following lesson plans, our hope is that educators can envision how these features have been thought through and will come to life in the delivery of the lesson plan. For example, wait time, which is a feature of the Interaction component, is not necessarily explained in the lesson plan itself, but it is assumed that an effective SIOP® teacher will ensure that all students will be provided the time they need to meet the objectives of the lesson. Feature # 25 (students engaged approximately 90% to 100% of the period) is also not explicitly stated, but, again, as you read the meaningful activities included in the lessons, you can conclude that students are engaged.

In order to help Tomas—as well as other educators facing the same dilemma—we have outlined in these chapters how to effectively incorporate techniques and activities into thoughtful, standards-based SIOP® lessons. We know well the realities and challenges that come with working with diverse groups of students, and it was also important for us to consider not only that our students are second language learners but also that they have different language levels and academic needs. Thus, we have also identified how the various techniques that are used to support the development of content and language can be modified for beginners, intermediate, and advanced high English learners within one lesson.

Math Techniques and Activities

In the following section you will see a variety of techniques described that may be used across grade levels, as indicated. Many are followed by a lesson plan that shows how the technique is embedded within a lesson. Note that the blackline masters (BLMs) referred to are found in Appendix C.

SIOP® Math Techniques and Activities: Lesson Preparation

Notice in the lesson that follows the activity description that there are a number of components represented, not just Lesson Preparation. However, all aspects of the lesson support and reinforce the objectives that were set during Lesson Preparation and are reviewed during the Wrap Up at the conclusion of the lesson.

This activity focuses on the lesson preparation component in that it assesses students' level of under-standing of the lesson's content and language objectives. As you know, content and language objectives drive instruction; even in inquiry-oriented lessons there are objectives or outcomes that the teacher has in mind when planning the lesson. Consider using the following activity in an upcoming lesson.

Number 1–3 for Self-Assessment of Objectives

COMPONENT: Preparation & Review/Assessment

Grade Levels: K–12
Grouping Configuration: Individual in whole class setting
Approximate Time Involved: A few minutes, once objectives are determined
Materials: Only what is needed to post objectives

Description

This quick activity can be used at the beginning and/or the end of a lesson. It gives students the opportunity to assess their own understanding and progress in meeting the content and language objectives explicitly and in an engaging way. It is a helpful reminder to teachers that the lesson they prepare will be assessed either explicitly or implicitly at the end of the period. The teacher also has the chance to assess the level of understanding that has occurred among the students and how the next lesson on the concept should be adjusted.

At the beginning of a lesson, the activity can be used to check prior knowledge when introducing objectives. Before moving into the lesson, have students rate their knowledge about the concept that is the focus for the new lesson. Read the lesson's objectives aloud and have students respond by holding up 1–3 fingers to indicate their level of knowledge.

    1 = I don't know anything about the objectives.
    2 = I know or think I know something about the objectives.
    3 = I know enough to teach the objectives.

At the end of a lesson, the teacher reviews the objectives and asks students to rate how well they believe they have met the objectives, again holding up the number of fingers that corresponds with their self-rating.

    1 = I didn't or couldn't meet the objective.
    2 = I didn't meet the objective completely, but made progress.
    3 = I met the objective.

Teach students what they should consider in determining their ratings. For example:

    How do I know I met the objective?
    What was the goal of the objective?

Depending on their ratings, the teacher can gauge if key concepts or language have to be reviewed or reinforced, or if students have mastered the concept and can move on to the next lesson.

Grade K–2 Math Activity Application:

In the lesson example (Analyzing and Describing Characteristics of Geometric Shapes), the activity is used both as a lesson starter to check prior knowledge and as a wrap-up.

Lesson Concept: Characteristics of Geometric Shapes

Content Objectives:

SW demonstrate knowledge of the characteristics of two- and three-dimensional geometric shapes by defining what a square, triangle, rhombus, trapezoid, and hexagon are on a 4-Corners vocabulary chart.

Language Objectives:

SW discuss comprehension of the characteristics of two- and three-dimensional geometric shapes by writing and explaining the characteristics of a square, triangle, rhombus, trapezoid, and hexagon using the frame:

One characteristic of a _____ is _____ .

The math teacher gives students the opportunity to consider the lesson and what they've learned about geometric shapes. After posting the objectives, she has the class evaluate what they already know using the 1–3 self-assessment technique. She notes that many students show one finger. At the end of the lesson, the class repeats the technique and she is pleased to see that most students show three fingers. She notes who did not and plans additional work for them the next day.

SIOP® LESSON PLAN

Key: SW = Students will; TW = Teacher will; HOTS = Higher Order Thinking Skills (questions and tasks)

Lesson: Describing characteristics of two- and three-dimensional shapes
Grades: Kindergarten – Second

NCTM Standard: Analyze characteristics and properties of two- and three-dimensional geometric shapes and develop mathematical arguments about geometric relationships

Expectation: recognize, name, build, draw, compare, and sort two- and three-dimensional shapes

Visuals & Resources:

Books "The Shape of Things" by Dayle Ann Dobbs or

"Shapes, Shapes, Shapes" by Tana Hoban

"4-Corners Vocabulary" from *99 Ideas and Activities for Teaching English Learners with The SIOP® Model*

What Do you Know About Geometric Shapes? BLM 1

Chart paper and markers, Pattern Blocks for each group, and blocks to show on overhead

Key Vocabulary:	General Frames:
square, triangle, rhombus, trapezoid, hexagon, characteristic	I noticed . . . Something that is the same is . . . Something that is different is . . .
HOTS (Higher order thinking questions or tasks):	**Specific Frames:**
Why is it important to know what shapes surround us?	One characteristic of a _____ is _____.

Connections to Prior Knowledge/ Provide Background Information:

TW display a rhombus, trapezoid, and hexagon. Students are asked to show 1 finger if they do not know the names of the shapes, two fingers if they know the name of at least one shape, and 3 fingers if they know the names of all three shapes.

TW read one of the books from above or one of his or her choice on shapes to introduce the topic. TW ask students where else they see shapes in or outside of the classroom. SW answer in small groups and then share with the whole group.

Content Objectives:	Meaningful Activities:	Review/Assessment:
SW demonstrate knowledge of the characteristics of two- and three-dimensional geometric shapes by defining what a square, triangle, rhombus, trapezoid, and hexagon are on a 4-Corners Vocabulary Chart.	TW remind students of the shapes that were identified in the story and the shapes they discussed in their groups in the Building Background component above.	
	TW distribute pattern block sets to each of the groups.	
	The set consists of a square, triangle, rhombus, trapezoid, and hexagon.	
	The teacher will ask students to take a few minutes in their groups to talk about the different shapes and answer the following questions:	
	What do you notice?	
	How are they the same?	
	How are they different?	
	SW share in their groups and answer using the general frames above.	Student responses
	I noticed . . . Something that is the same/different is . . . *(Please note, the teacher must explicitly teach and model the frame.)*	Student responses and appropriate use of the frame.
	TW then display each on the overhead and name each shape, explicitly identifying them as geometric shapes.	
	As the teacher displays and names each shape, SW share out their ideas of what they noticed about each shape in their groups.	
	TW explicitly teach the word *characteristic* and explain to students that what they are sharing are the characteristics of each of the shapes.	

(continued)

SIOP® LESSON PLAN *(continued)*

Content Objectives:	Meaningful Activities:	Review/Assessment:
	SW then be assigned one shape that they will use to complete a 4-Corners Vocabulary Chart for their group.	
	SW divide their piece of chart paper into four squares.	
	SW create their chart, including an illustration (representing the shape), a definition (describing the shape), a sentence (that includes the word) and the word itself.	
	SW present their completed chart to the group.	Completed 4-Corners Vocabulary Chart
	TW post charts on a math word wall in the room so that children can have them for reference.	Presentation of their charts

Language Objectives:

SW demonstrate comprehension of the characteristic of two- and three-dimensional geometric shapes by writing and explaining the characteristics of a square, triangle, rhombus, trapezoid, and hexagon using the frame:	TW provide students a sheet (BLM 1) that has each shape drawn and labeled.	
	TW model and guide students in practicing how to appropriately use the specific frame.	
One characteristic of a _____ is _____ .	Below each shape, SW individually complete the specific frame:	
	One characteristic of a _____ is _____ .	Student responses on BLM 1 and appropriate use of frames in writing and when orally sharing.
	TW remind students to refer to the 4-Corners Vocabulary Charts that they completed to help them complete their frames.	
	In pairs, SW share their completed frames.	Student responses in pairs.

Wrap-up: Rate your learning. TW ask students to think about how well they've learned the different shapes. SW show one finger if they feel they haven't learned; 2 fingers if they feel good about their learning but would like more information on shapes; and 3 fingers if they feel they have learned everything they need to about geometric shapes.

SIOP® Math Techniques and Activities: Building Background

In the lesson above, *Describing characteristics of two and three-dimensional shapes*, you can see the 4-Corners Vocabulary Chart used in a math lesson, grade band K-2. Notice how this activity provides oral language practice as students work together in groups and then present their chart to the class. In this way, new vocabulary terms are repeated and reinforced.

The 4-Corners Vocabulary Chart is a versatile activity that can be used with all grade levels and in a variety of grouping configurations. It can be seen in context in the preceding geometry lesson. As the following description states, one benefit of this activity is that it can be posted in the classroom to help build academic language and concepts throughout a lesson or unit. Following the description, we have shown how to differentiate this activity by language proficiency level.

4-Corners Vocabulary (adapted from 99 Ideas, p. 40)

SIOP SHELTERED INSTRUCTION OBSERVATION PROTOCOL

COMPONENT: Building Background

Grade Levels: All
Grouping Configuration: Partners, small group, whole class
Approximate time involved: 10–20 minutes
Materials: paper and pencil for upper grades, chart paper and marker for lower grades

Description

The purpose of a 4-Corners Vocabulary Chart is to have students contextualize key vocabulary words by creating a chart with an illustration of the word, the definition of the word, a sentence using the word in context, and the actual vocabulary word. The corners can also be adapted to include a synonym, antonym, what the word is not, etc.

Illustration	Sentence A triangle is a geometric shape.
Definition A triangle is three-sided figure: a two-dimensional geometric figure formed of three sides and three angles.	Vocabulary Word Triangle

For younger grades, the teacher might want to create the chart to save time; older students should have no problem creating their own. It is important that the teacher explicitly teach and model how to complete the 4-Corners Vocabulary Chart.

The charts should then be posted on the concept bulletin board so that students can continue to refer to them as they move through the topic being studied.

Application

Content Objective:

SW demonstrate knowledge of the characteristics of two- and three-dimensional geometric shapes by defining what a square, triangle, rhombus, trapezoid, and hexagon are on a 4-Corners Vocabulary Chart.

Language Objective:

SW demonstrate comprehension of the characteristics of two- and three-dimensional geometric shapes by writing and explaining the characteristics of a square, triangle, rhombus, trapezoid, and hexagon using the frame:

One characteristic of a _____ is _____ .

In the lesson example (Describing Characteristics of Two- and Three-Dimensional Shapes), students are being introduced to geometric shapes in order to describe characteristics of two- and three-dimensional shapes. In order to ensure that students will successfully move through the concept, it is important that they be introduced to the key vocabulary and given the opportunity to practice and apply it.

You will note in the lesson that students use the activity 4-Corners Vocabulary Chart, which is the focus of the content objective, to define the vocabulary and as an initial step to then focus on the standard, which is to describe the shape's characteristics (Language Objective).

Activity Differentiated Based on Levels of Language Proficiency

Beginning:

- Give students a preview of the academic language that they will need to understand for the lesson in their L1 (primary or native language). Include key content vocabulary and process/ functional vocabulary.
- Students are working in groups to complete the activity, so make sure that (when possible) you have a language broker (a student who speaks both the L1 of your ELs and English) in the group.
- On the sheet that students will be completing to meet the language objective, include responses that they can choose from to complete the frame: One characteristic of a _____ is _____ . They can also use illustrations to complete the frame.
- Consider allowing students to respond in L1 on the sheet as well as in their pairs when and if possible.

Intermediate:

- Give students a preview of the academic language that they will need to understand for the lesson in their L1 (primary or native language). Include key content vocabulary and process functional vocabulary.
- Have students compete all frames in writing and share at least two of the frames orally with their partners.
- Guide students in practicing the frames before they have to share.

(continued)

Advanced High:

- All of the modifications made for the Intermediate students would also be appropriate for this group.

- These students should be able to complete all frames on the sheet as well as share with their partners if the modeling outlined throughout the preceding lesson was consistent and ongoing.

The following technique is well known and quite effective for engaging students in their own learning.

KWL Chart (adapted from 99 Ideas, p. 27)

 COMPONENT: Building Background (Ogle, 1986)

Grade Level: All
Grouping Configurations: Individual, partners, small groups, whole class
Approximate Time Involved: 5 min–45 min depending on how it is used.
Materials: KWL Chart

Description

One of the best ways to activate prior knowledge and build background is the familiar KWL chart. The purpose of a KWL chart (What We **K**now/What We **W**ant to Learn/What We Have **L**earned) is to assess students' knowledge about a topic, uncover misconceptions, and, most importantly, allow students input about what they would like to learn about the topic.

As an alternative, add a box for "H" for "How We Find Out," before the "What We Have Learned" box. Students can use this new box to generate ideas for sources that can be researched to find additional information about the topic.

Grade 6–8 Math Activity Application

Lesson Concept: Circumferences of Circles

Content Objectives:
SW estimate and calculate the circumference of circles.

Language Objectives:
SW predict the circumference of a crop circle by using the following sentence frame:
   I think the circumferences of the crop circles are _____ feet and _____ feet.

1. Ask students to think quietly about what they know about crop circles. Give students approximately 1–2 minutes of quiet time.

2. Once all students are done thinking, have them work in groups or individually to brainstorm "What We Know" about crop circles. Provide students with a KWL chart and have them fill in the K column. Construct a KWL on chart paper, and record all information students brainstorm, even if it is inaccurate; clarification can take place when the third box is completed, after reading, learning about, and/or discussing the topic.

3. Build students' background knowledge of crop circles by having them view a short video on crop circles. Then have students fill out the column "What We Want to Learn" about crop circles. Gather students' information on the second column of the KWL chart. Inform students that they will learn how to find the distance around crop circles.

4. After students develop the formula for finding the circumference of a circle, have them use the formula to calculate the circumference of crop circles. At the end of the lesson, have students fill out the third box of the KWL Chart. Collect students' information and clarify any misconceptions during discussions. Post the KWL Chart so students can refer back to it as needed.

Activity Differentiated Based on Levels of Language Proficiency

The KWL Chart is a great way to access prior knowledge and explicitly connect past learning to new learning. However, we need to consider that some ELs have little or no prior knowledge about topics/terms such as crop circles. Therefore, instead of having ELs brainstorm on crop circles for the K portion of the KWL Chart, ask students to write about what they know about circles. After brainstorming about circles, build background on the topic of crop circles by showing students a short video. Now, all students are ready to brainstorm on "What We Want to Learn" about circles and crop circles.

Beginning:
Reinforce the question "What do you know about circles?" by drawing or showing a picture of a circle. Also, consider using the cognate *círculo* next to the word *circle* if the students' L1 is Spanish. Students at the beginning stages of second language acquisition might not be able to write in English what they know about circles. However in the KWL Chart, they can write about circles in their native language and draw pictures.

Intermediate:
Students at the intermediate level may also benefit from seeing a picture of a circle as they began to brainstorm about the topic. They can use the following sentence frame in the K column:

- Circles are . . .

Advanced High:
Students at this level can use the following sentence frame in the K column:

- What I remember about circles is . . .

The following lesson, *Circumferences of Circles*, demonstrates how a KWL Chart can be used effectively. Notice how many other components of the SIOP® Model are present in this lesson as well.

SIOP® LESSON PLAN *Circumferences of Circles*

Class/Subject Area(s): Math   **Grade Level:** 6–8
Unit/Theme: Circles   **Lesson Duration:** 90 minutes

NCTM Standards

Content	Process
Measurement: • Develop and use formulas to determine the circumference of circles and the area of triangles, parallelograms, trapezoids, and circles. Also develop strategies to find the area of more complex shapes.	☒ Problem Solving ☒ Reasoning & Proof ☒ Communication ☒ Connections ☒ Representations

Content Objective(s):
SW estimate and calculate the circumference of circles.

Language Objective(s):
SW predict the circumference of a crop circle by using the following sentence frame:
I think the circumferences of the crop circles are _____ feet and _____ feet.

Key Vocabulary:		Supplementary Materials:	
Content Vocabulary • Circle • Center • Circumference • Diameter • Radius • Crop circle	**Functional Vocabulary** • Predict	• KWL chart • Personal dictionaries • Chart paper • Variety of circular objects • Curly ribbon • Rulers • Measuring tape	• Scissors • Calculators • Markers • Chart paper • Lab sheet • Two-dimensional paper circle

SIOP® Features:

Preparation	Scaffolding	Grouping Options
X Adaptation of content	X Modeling	X Whole class
X Links to background	X Guided practice	__ Small groups
X Links to past learning	X Independent practice	X Partners
X Strategies incorporated	X Comprehensible input	X Independent

Integration of Processes	Application	Assessment
X Reading	X Hands-on	X Individual
X Writing	X Meaningful	X Group
X Speaking	X Linked to objectives	X Written
X Listening	X Promotes engagement	__ Oral

Lesson Sequence:

1. Activate students' prior knowledge of circles by having them complete a KWL Chart. Group students in triads, and ask them to think quietly for about 1 to 2 minutes of what they know about circles (ELs) or crop circles. Once time is up, provide each student with a KWL chart, and have students individually write and/or draw what they know about circles in the "What I Know" column. Once all students are done writing, have students share their thoughts with their group members. On chart paper, create a KWL Chart (What We Know, What We Want to Learn, What We Have Learned). In the "What We Know" column, gather all the information the groups brainstormed about circles.

2. Inform students that in this lesson they will find out more regarding circles by learning about crop circles. Ask students, "What is a crop circle?" To answer this question, have students participate in a Quickwrite by completing the following sentence frame: I think a crop circle is _____ . Collect students' responses.

(continued)

SIOP® LESSON PLAN *Circumferences of Circles* (continued)

Lesson Sequence:

3. To build students' background on the subject of crop circles, play video found on the Internet (search http://video.google.com/videosearch). After viewing the video, provide a quick explanation of what crop circles are. Ask students to reflect on what they would like to learn about circles and crop circles. Instruct them to write their thoughts in the KWL Chart under the "What I Want to Learn" column. Once all students are done writing, collect their data on the Chart's column titled "What We Want to Learn." (NOTE: Hopefully one of the things they want to know about crop circles is the distance around the circles. This will move the lesson into today's content and language objectives.)

4. Introduce content and language objectives by reading them aloud to students. Emphasize that today's lesson will focus on determining the distance around circles. In subsequent lessons, they will learn about how much area crop circles cover.

5. Frontload students with the lesson's key content vocabulary. Once you have provided a linguistic and nonlinguistic description of each term, provide each student with a two-dimensional circle. Tell them to pretend it is a crop circle. Ask the student to use a black marker to mark the center of the circle and to color the circumference red, the diameter green, and the radius blue. On the count of three, instruct students to compare their circle with that of a partner. If students have incorrect answers, they can use the back of the paper circle to mark the center and color the circumference, diameter, and radius. Have students add the terms to their Personal Dictionaries. (NOTE: For homework, students can take the circle home and decorate it to look like a crop circle.)

6. Show a picture of a crop circle. Tell students the larger crop circle has a diameter of 120 feet and the smaller one has a diameter of 60 feet. Use a measuring tape to model the length of 120 ft and 60 ft. Have students predict the circumference of the two crop circles. Tell students to write their predictions on a sticky note and place their note on the board. At the end of the period, students will compare their prediction to the actual circumferences of the crop circles. (Picture can be found at the following URL: http://ccdb.cropcircleresearch.com/info.cgi?d=ca2000ae&c=p)

7. In their triads, have students discuss how to determine the circumference of a circle. Write the groups' responses on the chart paper. Tell students that today they will find the circumference of circular objects by measuring the diameter and circumference with a ruler and/or measuring tape.

8. Model how to measure the diameter and circumference of a circular object using a ruler for the diameter and curly ribbon for the circumference. On the recording lab sheet, write the name of the circular object and the measurements of the diameter and the circumference.

9. Provide groups with a variety of circular objects, rulers/measuring tape, scissors, curly ribbon, and recording lab sheets. Have them measure the diameter and circumference of each circle and record their data on the recording lab sheet. Once they are done gathering their data, ask groups to compare the diameters to their circumferences. Do they notice any patterns? Have them discuss in their groups. Once all groups are done discussing, have a whole group discussion. (NOTE: The aim of this activity is for students to discover Pi.)

10. Explain to students what Pi is and how it is used to find the circumference of circles. Ask students, "If you want to find the circumference of any circle, what formula would you use?" To share their answers, have them participate in a Think, Pair, Share activity. Gather students' information. Formalize the formula for finding the circumference of a circle.

11. Show them a circular object. Tell them its diameter. Have them use their newly found formula to estimate the circumference of the object. Gather students' responses.

12. Show the picture of the crop circles. Ask students to determine the actual circumference of the circle crops using their newly discovered formula. Compare answers with their posted predictions. Have students complete the following sentence frame: When comparing my prediction to the actual answer, I concluded that . . .

13. To close today's lesson, revisit the KWL Chart. First review the content and language objectives, and then have students reflect on what they learned by filling out the "What I Learned" column of the KWL Chart. Gather students' reflections and fill out the chart paper's column "What We Learned."

Reflections:
After teaching the lesson, the teacher reflects on what worked, what did not work, and what revisions, additions, and/or deletions need to be made.

SIOP® Math Techniques and Activities: Comprehensible Input

For this component, in addition to the technique we describe, we direct your attention to the various lessons throughout the chapter because each one fosters comprehensible input. Notice that teachers model, use gestures, and show examples in the lessons—all good comprehensible input techniques.

The specific technique described here is one of many ways teachers make lessons comprehensible through visual representation.

Graphic Organizer: Math Representations

SIOP® COMPONENT: Comprehensible Input

Grade Level: 2–12
Grouping Configurations: Individual, partners, small groups, whole class
Approximate Time Involved: 10–20 minutes
Materials: Graphic organizers

Description:

Graphic organizers serve as a framework for students to learn and organize new content knowledge, making it comprehensible. With graphic organizers, students are able to represent a concept in multiple ways and begin to see connections within the various representations. The effectiveness of graphic organizers lies in the way they make abstract concepts more concrete and comprehensible. The graphic organizer presented in this technique (found in Appendix C, BLM 2) allows students to represent math topics concretely, pictorially, verbally, tabularly, graphically, and algebraically.

Grade 9–12 Math Activity Application

Lesson Concept: Analyzing Exponential Functions

Content Objectives:
SW: analyze an exponential problem situation.

Language Objectives:
SW: discuss in small groups the rate of change of an exponential function.

1. Provide students with one sheet of paper. Instruct them to fold the paper in half and then in half again. In the pictorial representation section of the graphic organizer, have students draw a picture of how many rectangles the paper represented with 0 folds, 1 fold, 2 folds, etc.

2. Instruct students to write a description of how many rectangles are created when the paper is folded 4 times. Have them write their description in the verbal representation section.

3. In the tabular section, have students fill out the table. The first column of the table represents the Number of Folds and the third column represents the Total Number of Rectangles. The middle column represents the process. The Process column allows students to discover and discuss what is happening mathematically.

4. Once students are done filling out the three columns of the table, instruct them to graph the table's information. Remind students to title the graph, label the axes appropriately, and determine reasonable intervals for the axes.

5. In the algebraic quadrant, provide students time to determine the function rule that represents the problem situation.

Activity Differentiated Based on Levels of Language Proficiency

The Math Representations Graphic Organizer, BLM 2, helps all English learners organize data in multiple ways. However, this graphic organizer is most helpful for students at early stages of language proficiency because data is presented in multiple representations, which allows them to comprehend content concepts without having to rely heavily on language.

Beginning:
For students at this level of language proficiency, the verbal section of the graphic organizer will be difficult to complete. Therefore, if possible, have them draw pictures in the verbal section.

Intermediate:
- Students at this level of language acquisition can complete the following cloze sentence for the verbal section of the graphic organizer.

  _____ folds equals _____ rectangles.

Advanced High:
- At this advanced level of language acquisition, students can complete the following cloze sentence.

  If a paper is folded _____ , _____ identical rectangles will be created.

In the following lesson, notice how the graphic organizer, *Math Representations*, is used so that students show math concepts in a variety of ways.

chapter 3 / Activities and Techniques for Planning SIOP® Mathematics Lessons

SIOP® LESSON PLAN *Bacteria Growth and Exponential Functions*

Class/Subject Area(s): Math
Unit/Theme: Exponential Functions

Grade Level: 9–12
Lesson Duration: 60 minutes

NCTM Standards

Content	Process
Algebra: • Generalize patterns using explicitly defined and recursively defined functions. • Understand relations and functions and select, convert flexibly among, and use various representations for them. • Understand and compare the properties of classes of functions, including exponential, polynomial, rational, logarithmic, and periodic functions.	☒ Problem Solving ☒ Reasoning & Proof ☒ Communication ☒ Connections ☒ Representations

Content Objective(s):
SW represent an exponential problem situation in multiple ways.

Language Objective(s):
SW discuss in small groups how bacteria grow.

Key Vocabulary:		Supplementary Materials:
Content Vocabulary • Exponential growth • Exponential function • E-coli bacteria	**Functional Vocabulary** • Represent	• Math representations graphic organizer (MRGO) • Clay • Bacteria growth video • Index cards

SIOP® Features:

Preparation
_X_ Adaptation of content
_X_ Links to background
_X_ Links to past learning
_X_ Strategies incorporated

Scaffolding
_X_ Modeling
_X_ Guided practice
_X_ Independent practice
_X_ Comprehensible input

Grouping Options
_X_ Whole class
_X_ Small groups
_X_ Partners
_X_ Independent

Integration of Processes
_X_ Reading
_X_ Writing
_X_ Speaking
_X_ Listening

Application
_X_ Hands-on
_X_ Meaningful
_X_ Linked to objectives
_X_ Promotes engagement

Assessment
_X_ Individual
_X_ Group
_X_ Written
_X_ Oral

Lesson Sequence:

1. Inform students that today's lesson will focus on investigating the growth of bacteria. Provide each student with an index card. On the index card have students answer the following question: "What do you know about bacteria?" To answer the question, instruct students to use the following sentence starter, "What I know about bacteria is . . . ." Once all students are done writing, have them share their responses by participating in a Think-Pair-Share activity. After sharing their responses with a partner, select two or three students to state to the whole group what they know about bacteria.

(continued)

SIOP® LESSON PLAN *Bacteria Growth and Exponential Functions* *(continued)*

Lesson Sequence:

2. In small groups, have students discuss how bacteria grow. Once all groups have finished discussing, select two groups to share what they discussed. If students do not know how bacteria grow, build background by showing them a video on bacteria growth (a video on bacteria growth can be found at the following website: http://www.youtube.com/watch?v=gEwzDydciWc&feature=related).

3. Introduce the lesson's content and language objectives.

4. Tell students they will use clay to model the growth of e-coli bacteria. Explain that an e-coli bacterium splits into two daughter cells every 20 minutes. Instruct students to use a piece of clay to represent one e-coli bacterium. Tell them to pretend 20 minutes have elapsed and the e-coli bacterium has divided into two daughter cells. Have students create the two daughter cells. Call this stage of growth Phase 1. Ask students to create the total number of e-coli bacteria in Phase 2. Phase 3? Phase 4? (For information on the growth rate and generation time of e-coli bacteria visit the following website, http://www.textbookofbacteriology.net/growth.html.)

5. Now that students have represented the growth of e-coli bacteria concretely, have them use the Math Representations Graphic Organizer (MRGO) (BLM 2) to represent Phase 5 and Phase 6 pictorially. (Teacher models using BLM 2, MRGO) Once students are done drawing Phase 5 and 6 on the graphic organizer, have them organize their data in the table section of the MRGO. The first column will represent the phase number and the third column will represent the number of bacteria. Ask students to determine the number of bacteria starting with Phase 0 and ending with Phase 9. Instruct students to use the middle column to discover and discuss what is happening symbolically. Have students determine how many bacteria there will be in Phase 30 and Phase X.

6. In the verbal/written section of the MRGO, have students identify the independent variable and dependent variable. In the function rule section, instruct students to write the function rule of this problem situation.

7. Following the teacher's modeling, have students graph the table's data. Remind students to title the graph, label the axes appropriately, and determine reasonable intervals for the axes. In the verbal/written quadrant, tell students to write a reasonable domain and range for this problem situation.

8. In small groups, have students discuss how the bacteria grew. Was the rate of change constant? If the answer is no, how did the bacteria grow?

9. At this time, introduce the concept of exponential growth and explain the characteristics of an exponential function. Have students process the meaning of exponential growth and exponential function by adding these two terms to their Personal Dictionaries.

10. As a final wrap-up activity, have students participate in Outcome Sentences: "I think . . ." "I wonder . . ." "I learned . . ." "I liked . . ." "I discovered . . ."

11. Review the lesson's content and language objectives with students.

Reflections:
After teaching the lesson, the teacher reflects on what worked, what did not work, and what revisions, additions, and/or deletions need to be made.

Concluding Thoughts

In this chapter you have seen a variety of techniques and activities presented, most of which were shown with an example of how they can be used within a math lesson. It is important that you keep in mind that although they are presented by SIOP® components for organizational purposes, the techniques are intended to support the lesson's language and content objectives, and not to be isolated activities. Our research has shown that lessons are effective for English learners when all of the components of the SIOP® Model are evident in the lesson, and that lessons are ineffective when teachers pick and choose only their favored techniques.

Let's revisit Tomas, the teacher described at the beginning of the chapter. In anticipation of his principal's visit, Tomas developed a lesson plan whose content was driven by the lesson's objectives. Once he decided on the specific objectives students would meet during the math period, he was then able to select activities that would assist and support students in meeting the lesson's objectives. Tomas began the lesson that his principal observed by stating the content and language objectives for students, and then he maintained the lesson's focus on the objectives. Students were engaged and participated in an activity that reinforced the learning. At the conclusion of the lesson, Tomas reviewed the objectives and used the technique "Number 1–3 for Self-Assessment of Objectives." Both he and his principal were pleased to see that most of the class held up three fingers, indicating that they knew the objectives well enough to teach them to someone else.

In the next chapter you will read about more techniques, those for the SIOP® components Strategies, Interaction, Practice & Application, and Review & Assessment.

Activities and Techniques for Planning SIOP® Mathematics Lessons

Strategies, Interaction, Practice & Application, Review & Assessment

Melissa Castillo, Araceli Avila and Jana Echevarría

In this chapter, as in Chapter 3, we have organized a variety of proven techniques and activities around the components of the SIOP® Model. You will find ideas for the components Strategies, Interaction, Practice &Application, and Review &Assessment. Again, remember that it is important to consider the components together as a whole, rather than picking only certain ones to include in lessons. As you see in the lesson plans, although one technique or activity is highlighted, there are other aspects of the lesson that provide opportunities for students to build background, apply learner strategies, answer higher order questions, use hands-on materials and manipulatives, interact with the teacher and other students, review and assess key math concepts and vocabulary, support both the content and language objectives, and use a variety of techniques to make the content comprehensible.

Please keep in mind that each of the techniques presented in this chapter may be used across various grade levels, K–12; they are not restricted to the grade level we show in the sample lessons. At the top of each technique description, the optimal grades in which the technique may be used is stated. Also, the lesson plan following the technique is specific to a grade band, but you might imagine how the technique would be used in another grade level to support the objectives of a particular lesson. Note that the blackline Masters (BLMs) referred to in the chapter are found in Appendix C.

If you haven't already done so, it is recommended that you read the introduction in Chapter 3 which provides an important context for the information presented in Chapters 3 and 4.

SIOP® Math Techniques and Activities: Strategies

The technique "You Are the Teacher!" is a fun way to get students engaged in the lesson. The sentence frames scaffold students' oral expression. Although it is shown in a lesson suited for young children, this technique is popular with older students as well.

You Are the Teacher! (adapted from *99 Ideas*, p. 92)

SIOP® COMPONENT: Strategies

Grade Levels: 1–12
Grouping Configurations: Small groups and whole class
Approximate Time Involved: 45–60 minutes (less for younger learners)
Materials: Chart paper, markers, and text

Description

"You Are the Teacher" puts students in the role of the teacher, allowing them to teach and learn from one another. After students have been taught a math process, concept, or solution to a word problem, they are given the opportunity to practice it in small groups. Then each group organizes what they have learned on a piece of chart paper. They create a "mini" presentation of their work in their small groups to teach to the other members of the class.

Each group sets up at a station in the room. For example, if the class is divided into four groups, there will be four stations. In order to determine who will stay and who will rotate, have students count off or letter off in their small groups. At the teacher's cue (he or she will call a letter or number), one group member from each station will stay and present while the others in the group move clockwise to the next station. The teacher gives students the cue for each rotation. Depending on the age group and the amount of information being presented, determine the appropriate amount of time needed before each rotation. For young learners and English learners, consider having two students work together as "teacher" to support one another.

Rotate until students end up back at their original station.

In order for this activity to be applied effectively, the teacher must explicitly teach the process and give students the opportunity to practice it several times.

Grade K–2 Math Activity Application

Lesson Concept: Standard and Nonstandard Units of Measurement

Content Objective:
SW demonstrate comprehension of using nonstandard and standard units of measurement by comparing measurements of nonstandard (hands and feet) and standard (ruler) units on a graphic organizer.

Language Objective:

SW demonstrate evaluation of using nonstandard and standard units of measurement by explaining which unit of measurement is most accurate and orally sharing their measurements in a "You Are the Teacher" activity.

In the lesson example (Using Nonstandard and Standard Units of Measurement), "You Are the Teacher" is a meaningful activity that is part of the Language Objective. Through the activity, students explain, using sentence frames, the units of measurement they have learned. In math classes, this activity reinforces the math concept and also provides valuable oral language experience using academic language.

Note that in the lesson, students are given a specific way to organize and prepare what they have learned in order to present it to the rest of the class. Also, the teacher needs to make sure that information on the poster is accurate so that it can be correctly "taught" by students.

Activity Differentiated Based on Levels of Language Proficiency

Beginning:
- Give students a preview in their native language (L1) of the academic language that they will need in order to understand the lesson in English. Include key content vocabulary and process/functional vocabulary.
- As students are working in groups to complete the activity, make sure that (when possible) you have a language broker (a student who speaks both the L1 of your ELs and English) in the group.
- As students are working on their presentations, the teacher should also model the sentence frames being used and guide beginners in practice using frames.
- When students rotate to present to the group, do not leave the Beginning level student as a presenter.

Intermediate:
- Give students a preview of the academic language that they will need to understand for the lesson in their native language. Include key content vocabulary and process functional vocabulary.
- As students are working on their presentations, the teacher should also model the sentence frames being used and guide beginners in practice using frames.

(continued)

- As students rotate to present to the group, make sure they are paired for the presentation and give the Intermediate students only a portion of the information to present.

Advanced High:

- All of the modifications made for the Intermediate students would also be appropriate for this group.

- These students should also be ready to present on their own if the modeling outlined throughout the lesson was consistent and ongoing.

As you can see in the following lesson, the technique "You Are the Teacher!" is embedded in order to support the content and language objectives and to scaffold students' active participation.

SIOP® LESSON PLAN

Key: SW = Students will; TW = Teacher will; HOTS = Higher Order Thinking Skills (questions and tasks)

Lesson: Using Nonstandard and Standard Units of Measurement
Approximate Duration: 60 minutes
Grades: Kindergarten–Second

NCTM Standard: Understand measurable attributes of objects and the units, systems, and processes of measurement.

Expectation: Understand how to measure using nonstandard and standard units.

Visuals & Resources:

Book, "How Big Is a Foot?" by Rolf Miller

BLM 3 Let's Measure

Rulers

"You Are the Teacher!" from *99 Ideas and Activities for Teaching English Learners with the SIOP® Model*

Key Vocabulary:	General Frames:
nonstandard, standard, measurement, foot, feet, inches	I can use . . .
HOTS (Higher order thinking questions or tasks): Which unit of measurement provides the most accurate measurement? Why?	**Specific Frames:** The best way to measure is using a _____ because _____.

Connections to Prior Knowledge/ Provide Background Information:
TW ask students if they have ever wanted to know how long, tall, or wide something was. How did you figure it out? Why was it important?
Read the Story "How Big Is a Foot?"

(continued)

SIOP® LESSON PLAN *(continued)*

Content Objectives:	Meaningful Activities:	Review/Assessment:
1. SW demonstrate comprehension of using nonstandard and standard units of measurement by comparing measurements of nonstandard (hands and feet) and standard (ruler) units on a graphic organizer.	• TW read the book "How Big Is a Foot?" • TW ask questions throughout the story about why the apprentice in the book is unable to make a bed that fits the queen. • SW answer and share out in small groups. • TW ask students what they might have to use for measurement besides feet, as the apprentice did in the story. • SW answer in groups using the general frame.	Student responses
	I can use . . . (Teacher must explicitly teach the frame, model, and guide students in practice.)	Teacher observation of student responses
	• TW then explain to students that they have tools they can use to measure such as their own hands, feet, and fingers. • TW also explicitly teach the vocabulary term *nonstandard*, making sure students understand that these tools are nonstandard units used to measure. • TW explain that they will be using these tools to measure and model several examples. • In small groups, SW use their nonstandard units (hands, feet, fingers) to measure items such as a desk, length of the chalkboard, and a book. Note: Since students are working in groups, ask each of them to put their name next to their measurement on BLM 3. • SW list their measurement on chart paper using the format of the graphic organizer (BLM 3). • TW bring students back together and ask them if there are any other tools they can use or that the apprentice could have used that aren't part of them.	Student responses and appropriate use of the frame Measurements on chart paper and completed graphic organizers
	• SW respond in small groups and share out using the general frame: I can use. . . • TW share with them or remind them, depending on the grade level, about the ruler (standard unit) and how we can use it to measure in inches and feet. TW also explicitly teach the vocabulary *standard*. • TW model measuring with the ruler the same examples she used for *nonstandard*. • SW then be asked to measure with the ruler the same objects as in the earlier example. • In their groups SW now add to their graph on the chart paper what the new measurements are using the ruler.	Student responses and appropriate use of the general frame Student measurements on graphic organizer

Language Objectives:	Meaningful Activities:	Review/Assessment:
1. SW demonstrate evaluation of using nonstandard and standard units of measurement by explaining which unit of measurement is most accurate and orally sharing their measurements in a "You Are the Teacher!" activity.	• TW ask students if they notice a difference in the measurement results that they have listed. • SW respond in groups and share out. • TW also ask them if there is one way to measure that might be better than another, and why? • SW answer in their groups and share out. • Students should come to the conclusion that nonstandard units of measurement are less reliable than standard units of measurement. TW guide discussion using prompts and questions to elicit students' thinking. • In their groups, SW complete the following specific frame at the bottom of their chart: The best way to measure is using a _____ because. . . . • In their groups, SW then set up their completed graphs at a designated station in the room. For example, if there are four groups, there will be four stations, one in each corner. • SW leave one of their peers at their station and rotate to the nearest station clockwise. • The student who stays at the station will present the poster and explain which measurement was most accurate and why. Students will continue to rotate until they return to their original station.	Student responses Student responses and appropriate use of the specific frame Student presentations

Wrap-up: TW ask students to answer the following questions on a note card:

What is an example of a nonstandard unit of measurement?

What is an example of a standard unit of measurement?

SW use the frames: A nonstandard example is _____. A standard example is _____.

Another technique that supports the Strategies component is "Vocabulary Alive." Think about ways that you might use this technique in your own math class.

Vocabulary Alive (Created by Cristina Ferrari, Brownsville ISD)

COMPONENT: Strategies

Grade Levels: All
Grouping Configurations: Individual, partners, small groups, whole class
Approximate Time Involved: 5–15 minutes, depending on how it is used
Materials: Vocabulary words on index cards

Description

"Vocabulary Alive" is a memory system that involves acting out key vocabulary. This promotes strategic thinking in that students have to conceptualize and act out the meaning of math vocabulary. This technique boosts the self-confidence of those ELs who do not have the English proficiency to express themselves clearly. Furthermore, students at various levels of language proficiency find "Voca-bulary Alive" engaging because it gets them out of their chairs and allows them to move around. Effective use of "Vocabulary Alive" depends on students' understanding of the meaning of the words used for the activity. Therefore, students need to define and internalize the meaning of the key words prior to participating in "Vocabulary Alive."

Grade 9–12 Math Activity Application

Lesson Concept: Translating Parabolas on a Coordinate Plane

Content Objectives:
SW investigate the effects of changes in c on the graph of $y = x^2 + c$.

Language Objectives:
SW act out vocabulary words by participating in "Vocabulary Alive."

The word is _____ and it looks like this _____.

1. Write the words *translation, vertical axis, horizontal axis, origin, coordinate plane, quadrants, quadratic parent function, parabola,* and *vertex* on index cards.
2. Group students into nine small groups.
3. Provide each group with a vocabulary word.

(continued)

4. Instruct students to use their Personal Dictionaries or 4-Corners Vocabulary Chart to write a definition of the assigned word on the back of the index card and decide on how to act out the word.

5. Ask students to form a circle around the perimeter of the room. Students must stay with their assigned group.

6. Number groups off 1 through 9.

7. Group 1 acts out a word by saying: "The word is *translation* and it looks like this. . ."

8. Teacher and whole class act out the word by saying: "The word is *translation* and it looks like this. . ."

9. Group 2 acts out a word by saying: "The word is *vertical axis* and it looks like this. . ."

10. The teacher and the whole class act out the word by saying: "The word is *vertical axis* and it looks like this. . ."

11. The teacher and the whole class go back to words 1 and 2 and repeat the action by saying: "The word is *translation* and it looks like this. . .; The word is *vertical axis* and it looks like this. . ."

12. Repeat this process until all words and actions are shared and repeated.

13. At the end of the activity, the teacher can assess student comprehension by acting out the word and having students state what word was acted out.

Activity Differentiated Based on Levels of Language Proficiency

"Vocabulary Alive" works effectively with children and adults regardless of varying levels of language proficiency. However, consider the following accommodations based on levels of language proficiency.

Beginning:

- To accommodate the linguistic needs of beginning English learners, consider writing the vocabulary words in L1 and L2 and providing a nonlinguistic representation.

- Instead of having groups come to consensus on how to act the word out, the teacher acts out the word and students repeat the word and action. The following sentence frame can be used: " _____ and it looks like. . ."

Intermediate:

- To accommodate the linguistic needs of intermediate English learners, consider writing the vocabulary words in L1 and L2 and providing a nonlinguistic representation.

- The teacher models how to write a definition of a vocabulary word by doing a Think-Aloud and acts out a vocabulary word by using the sentence frame: "The word is _____ and it looks like this. . ." Students then work in groups to come to consensus on how to act out the remaining words.

Advanced High:

- The teacher models one vocabulary word by acting it out and using the sentence frame: "The word is _____ and it looks like this. . ." Students then work in groups to come to consensus on how to act out the remaining words.

The following lesson illustrates how the technique "Vocabulary Alive" can be used in a math lesson. Although this lesson is for high school students, the technique can be used at all grade levels, as mentioned previously.

SIOP® LESSON PLAN: *Translating Quadratic Functions*

Class/Subject Area(s): Math   **Grade Level:** 9–12
Unit/Theme: Quadratic Functions   **Lesson Duration:** 90 minutes

NCTM Standards

Content	*Process*
Algebra: • Understand and perform transformations such as arithmetically combining, composing, and inverting commonly used functions, using technology to perform such operations on more-complicated symbolic expressions. **Geometry:** • Understand and represent translations, reflections, rotations, and dilations of objects in the plane by using sketches, coordinates, vectors, function notation, and matrices.	☐ Problem Solving ☒ Reasoning & Proof ☒ Communication ☒ Connections ☒ Representations

Content Objective(s):
SW investigate the effects of changes in c on the graph of $y = x^2 + c$.

Language Objective(s):
SW act out vocabulary words by participating in "Vocabulary Alive."
The word is _____ and it looks like this _____ .

Key Vocabulary:		*Supplementary Materials:*
Content Vocabulary • Translation • Vertical axis • Horizontal axis • Origin • Coordinate plane • Quadrants • Quadratic parent function • Parabola • Vertex	**Functional Vocabulary** • Compare • Contrast	• 10′ by 10′ coordinate grid on classroom floor • Ordered pair cards (Domain: $-10 \geq x \leq 10$; Range: $-10 \geq y \leq 10$) which represent vertically translated parabolas. • Math representations graphic organizer (MRGO) • Patty paper • Masking tape • Ordered pair and quadratic function cards • Response boards and expo markers • Graphing calculators

SIOP® Features:

Preparation	**Scaffolding**	**Grouping Options**
X Adaptation of content	X Modeling	X Whole class
X Links to background	X Guided practice	X Small groups
X Links to past learning	X Independent practice	X Partners
X Strategies incorporated	X Comprehensible input	X Independent

Integration of Processes	Application	Assessment
✗ Reading	✗ Hands-on	✗ Individual
✗ Writing	✗ Meaningful	✗ Group
✗ Speaking	✗ Linked to objectives	✗ Written
✗ Listening	✗ Promotes engagement	__ Oral

Lesson Sequence:

1. Activate students' prior knowledge by having them participate in a Graffiti Write: Give students two minutes to provide linguistic and nonlinguistic representations of what the word DANCE means to them. As students are writing/drawing, walk around the room to determine what they know about the word DANCE. Once time is up, select a couple of students to share their thoughts.

2. Inform students that there are many different types of dances around the world. Explain that in the United States line dances are very popular. Ask students to give you thumbs up if they know the meaning of *line dance*, thumbs down if they do not know the meaning of *line dance*, or thumbs in the middle if they have some idea of what *line dance* means. If the majority of your students give you thumbs in the middle or thumbs down, proceed to step 3. If the majority of your students give you thumbs up, ask one or two students to explain the meaning of a line dance. Have students line dance to the Cupid Shuffle and proceed to step 4.

3. For many of our recent immigrant students, a line dance is a new concept. Therefore, build background by playing a portion of the song, Cupid Shuffle. Inform students that the Cupid Shuffle is a type of line dance. Have students view a video of students dancing the Cupid Shuffle. After they view the video, have students line dance to the Cupid Shuffle and proceed to step 4.

4. Tell students that in the Cupid Shuffle line dance, a person moves 4 steps to the right, 4 steps to the left, kicks 4 times, turns 90 degrees counterclockwise, and starts the pattern all over again. Ask students if they remember from middle school the math term that describes moving to the right, left, up or down. Once they share their responses, ask them if they remember the math term for turning. Connect the words *translate* and *rotate* to the Cupid Shuffle steps.

5. Inform students that today's lesson will deal with translating the parent function $y = x^2$ on a coordinate plane. Introduce the content and language objectives.

6. Review the lesson's key vocabulary by providing a brief linguistic and nonlinguistic explanation of the terms. Once you are done with the explanations, have students complete the 4-Corners Vocabulary worksheet for each word. Once students finish the 4-Corners Vocabulary activity, have them participate in "Vocabulary Alive." For each word, have students use the sentence frame: "The word is _____ and it looks like this _____ ."

7. Use a graphing calculator or patty paper to model translating $y = x^2$ on a coordinate plane. Compare and contrast the change in size, orientation, and position of the pre-image and image. How are the x and y values changing? How did the parent function change? What is the function rule of the image? Use the Math Representation Graphic Organizer to collect data. Have students draw a picture, describe verbally how the graph of the parent function changed, record the changes of the x and y values on the table, graph the table's data, and write the new function rule. Guide students through one or two more examples. Then let them work in pairs on one or two more problems. Students may use the graphing calculator or patty paper to translate the parabolas.

8. With masking tape, construct a 10 feet by 10 feet coordinate plane on the floor. Model how to graph a quadratic parent function on the 10′ by 10′ coordinate plane. Once the parent function is graphed, model how to translate it 3 units up. Ask the following questions:

   - How did the parent function change? Did it move left, right, down or up?
   - What is the position of the translated parabola?
   - After the translation, did the size of the parabola change? If yes, how?
   - After the translation, did the orientation change? If yes, how?

9. Group students in sets of five. Provide each group with six cards and five respond boards. Five cards will have one of the following ordered pairs: (0, 0), (1, 1), (−1, 1), (2, 4), (−2, 4). The sixth card will have an equation that describes the vertical translation of the parent function. (For example: $y = x^2 + 3$). Explain that each group's responsibility will be to graph themselves on the coordinate plane and then translate the parabola the number of units reflected on their equation. Give groups

(continued)

SIOP® LESSON PLAN: *Translating Quadratic Functions* *(continued)*

three to four minutes to discuss how they will model the translation of their assigned parabola to the rest of the class.

10. Instruct group one to graph themselves on the coordinate grid based on the ordered pairs. Once they have graphed the quadratic parent function, have them translate.

11. After each translation, the students who are sitting write the new function rule on their respond boards. Have students compare their findings. Repeat steps 10 and 11 until all groups are done.

12. Have a whole class discussion to determine if congruence (same size, same shape) depends on position. Reinforce the meaning of *translation*.

13. To assess what students learned about translating, show them graphs of translated quadratic parent functions. Have students determine the equation of the graphed function by writing the answer on their response boards.

14. Wrap up today's lesson by reviewing the content and language objectives.

Reflections:
After teaching the lesson, the teacher reflects on what worked, what did not work, and what revisions, additions, and/or deletions need to be made.

SIOP® Math Techniques and Activities: Interaction

Next we present three activities or techniques for the component, Interaction. As English learners develop proficiency in English, it is critical that they have opportunities to improve their oral language skills, which is done through oral practice.

As its name implies, the "Conga Line" technique places students in two lines facing one another to facilitate interaction. It is also called "Inside-Outside Circle" when two concentric circles of students face each other.

Conga Line (adapted from *99 Ideas*, p. 110)

COMPONENT: Interaction

Grade Levels: All
Grouping Configurations: Partners, small groups, whole class
Approximate Time Involved: 15–60 minutes, depending on how it is used
Materials: Information for students to share orally (written information, pictures, illustrations, white boards, etc.)

Description

The "Conga Line" is a kinesthetic activity that involves all students in a class and provides them an opportunity to practice key content concepts and develop oral language. This activity gives English learners the benefit of repetition and varied use of language. It can be used in any content area, but is illustrated here with math. Follow these easy steps to implement the "Conga Line" activity in your classroom:

1. Students are directed to perform a task.

2. Students group themselves in pairs and number off as 1 or 2. For a predetermined amount of time, student 1 shares his response with student 2. After time is up, the students reverse roles.

3. Instruct all 1's to form a line.

4. Ask all 2's to find their partner and stand in front of them.

5. Two lines are now formed. Line 1 will be stationary and line 2 will move.

6. Ask the student in the farthest left corner of line 2 to run down the Conga Line and join the end of line 2. All students in line 2 will move one person to the left in order to make space for the student who ran down the Conga Line. (Students can dance as they go down the Conga Line, or they can run down the Conga Line while their classmates high-five them.)

7. Students share with their new partner. When pairs are done sharing, they can give a signal to indicate they are done (e.g., thumbs up or high-five each other). Once all pairs are finished, the teacher gives a signal (e.g., plays music or rings a bell) and line 1 stays in place while the next student in the farthest left corner of line 2 runs down the Conga Line.

8. Repeat steps 6 and 7 at least four or five more times.

Grade 6–8 Math Activity Application

Lesson Concept: Comparing and Ordering Fractions, Decimals, and Percents

Content Objectives:
SW compare and order fractions, decimals, and percents on a number line.

Language Objectives:
SW write two equivalent numeric representations of a fraction, decimal, or percent.

_____ (fill in with fraction, decimal or percent) is equivalent to _____ and to _____.

SW use the words _greater than, less than,_ or _equal_ to compare fractions/decimals/percents while participating in a "Conga Line" activity.

My _____ (fraction, decimal, percent) is _____ (greater than, less than or equal to) yours.

Lesson Sequence:
1. Students are given a fraction, decimal, or percent card. Their task is to compare their fraction/decimal/percent with that of another student and determine who has the greatest number. Before students compare their number with that of a partner, instruct them to individually convert their fraction, decimal, or percent to the other two representations. Once they are done, ask them to complete the following sentence frame:

_____ is equivalent to _____ and to _____ .
fraction, decimal, percent

2. Students group themselves in pairs and number off as 1 or 2. For a predetermined amount of time, student 1 reads his sentence frame to student 2. After time is up, students reverse roles. Once they are finished sharing, they determine who has the greatest number and complete the following sentence frame and number sentence. The student with the greatest fraction/decimal/percent immediately raises her hand.

My _____ is _____ yours.

     fraction, decimal, percent   greater than, less than or equal to

_____   _____   _____

write fraction, decimal, percent   >, <, =   write fraction, decimal, percent

3. Instruct all 1's to form a line.

4. Ask all 2's to find their partner and stand in front of them.

5. Two lines are now formed. Line 1 will be stationary and line 2 will move.

6. Ask the student in the farthest left corner of line 2 to run down the Conga Line and join the end of line 2. All students in line 2 will move one person to the left in order to make space for the student who ran down the Conga Line. (Students can dance as they go down the Conga Line, or they can run down the Conga Line while their classmates high-five them.)

7. Students share with their new partner. When pairs are done sharing, the student with the greatest fraction/decimal/percent raises his hand. Once all pairs are finished, the teacher gives a signal (e.g., plays music or rings a bell) and line 1 stays in place while the next student in the farthest left corner of line 2 runs down the Conga Line.

8. Repeat steps 6 and 7 at least four or five more times.

Activity Differentiated Based on Levels of Language Proficiency

In spite of different levels of language proficiency, all students can participate in the "Conga Line" activity. To ensure success for all students, differentiate the sentence frames according to language proficiency levels.

Beginning:

- _____ equals to _____ and _____ . (Student uses numeric values in blanks.)
- _____ is _____ (greater than, less than, equal to) _____. (Student uses numeric values in blanks.)

Intermediate:

- _____ is equivalent to _____ and to _____. (Student uses numeric values in blanks.)
- My _____ (fraction, decimal, percent) is (greater than, less than, equal to) yours.

Advanced High:

- _____ is equivalent to _____ and to _____. (Student fills in the blanks with the word that represents the number.)
- My _____ (fraction, decimal, percent) is (greater than, less than, equal to) yours because _____.

The following technique promotes interaction and also provides opportunities for students to practice and apply concepts (Practice & Application), use higher level thinking (Strategies), and demonstrate what they have learned (Review & Assessment).

Group Responses with a White Board

(adapted from *99 Ideas*, page 107)

 COMPONENT: Interaction

Grade Levels: All
Grouping Configurations: Individual, partners, small group, whole class
Approximate Time Involved: Varies, depending on how it is used
Materials: Individual white boards, white board markers

Description

Students can work individually, with a partner, or in their small groups.

The activity can begin with partners and small groups and move toward individual student work. Each student has his or her individual white board, but students are allowed to formulate responses together. Students will sometimes be asked to share their responses on their individual boards or together on one white board as a small group.

The teacher asks questions for the students to answer or gives them problems to solve. In the beginning, students are allowed to work with partners. After allowing sufficient wait time, the teacher allows partners to then share with their small groups of four. The group decides which is the best answer and one group member holds up the white board. The teacher might also ask students to write down the same answer on their white boards so the whole group can share.

As students are developing an understanding of the concept being targeted, it is important that they be asked to formulate responses individually and share them as well. The teacher can then assess how well the concept is being mastered by each student.

Grade K–2 Math Activity Application

Lesson Concept: Numbers and Operations: Addition and Subtraction

Content Objective:
SW demonstrate application of adding and subtracting whole numbers by solving equations (addition and subtraction) on white boards.

Language Objective:
SW demonstrate analysis of adding and subtracting whole numbers by writing addition and subtraction sentences and orally sharing with a partner.

Lesson Sequence:

In the lesson example (Numbers and Operations: Addition and Subtraction), "Group Response with a White Board" is the activity that students are participating in to complete problem solving and then check their work with a partner. It is also an opportunity for students to scaffold the procedure of formulating and solving problems using white boards, as well as demonstrating their understanding of the concept before they move on to the Language Objective of this particular lesson.

The teacher is able to assess students' level of understanding and decide whether students can transition from using numbers and symbols to solving word problems.

Activity Differentiated Based on Levels of Language Proficiency

Beginning:

- Give students a preview in their L1 of the academic language that they will need to understand the lesson. Include key content vocabulary and process/functional vocabulary.
- Allow students to respond to addition and subtraction problems in their L1. The teacher can determine first if the students understand the concept.
- As students share in pairs, allow them to share in their L1.
- As they work in pairs, model and guide students in using the L2 to share their responses.
- The teacher and students have modeled for one another throughout the lesson; at this point, determine through checking for understanding if it is enough. The teacher may need to model several examples before group and independent practice occurs.

Intermediate:

- Give students a preview in their L1 of the academic language that they will need to understand the lesson. Include key content vocabulary and process/functional vocabulary.
- As students share in pairs, have them share in L1 and L2, to give them practice in developing proficiency in the concept as well as proficiency in the target language (English).
- In the beginning, students should also be allowed to plug numbers in the frames instead of vocabulary words, and then move to the words. For example:

    _____2_ plus _____2_ equals _____ 4.

Advanced High:

- All of the modifications made for the Intermediate students would also be appropriate for this group.
- Students in this group need to be given multiple opportunities to practice the frames in guided practice in the whole class, their small group, and in their pairs.

In the lesson that follows, notice how effectively the technique, "Group Response with a White Board" is used. Also note the level of oral language practice that occurs as students share their answers with partners.

SIOP® LESSON PLAN

Key: SW = Students will; TW = Teacher will; HOTS = Higher Order Thinking Skills (questions and tasks)

Lesson: Numbers and Operations: Addition and Subtraction
Approximate Duration: 30 minutes
Grades: Kindergarten – Second

NCTM Standard: Understand meanings of operations and how they relate to one another.

Expectation: Understand the effects of adding and subtracting whole numbers.

Visuals & Resources: "The M&M's Addition Book" by Barbara McGrath
Snack M&M bags (1 for each student), white boards, dry erase markers, and tissue paper

Overhead

"Group Response with a White Board" from *99 Ideas and Activities for Teaching English Learners with the SIOP® Model*

Key Vocabulary:	General Frames:
addition, subtraction, plus, minus, equals	**Specific Frames:**
HOTS: How can you turn an addition problem into a subtraction problem?	_____ plus _____ equals _____ .
Why are both problems important?	_____ minus _____ equals _____ .

Connections to Prior Knowledge/ Provide Background Information:
TW explicitly review the concept of adding and subtracting single- and double-digit numbers. TW review key vocabulary.
TW ask students when they add and subtract in school and outside of school.
SW answer in their small groups and then share out their ideas.

Content Objectives:	Meaningful Activities:	Review/Assessment:
1. SW demonstrate application of adding and subtracting whole numbers by solving equations (addition and subtraction) on white boards.	• TW read the book "The M&M's Addition Book" • As the teacher reads, he or she will stop and ask students to solve addition problems being read using their M&Ms. • TW also ask them to consider and solve what the subtraction problem could be from the story. • SW solve in pairs and share their solutions with opposite pairs in their groups of four. • TW have students model on the overhead so that students can assess whether they have the correct number of M&M's. • TW then ask students to write and solve equations on their white boards, for example: 3 + 5 = 8. • TW also ask students to solve subtraction problems. • SW write their equations on their white boards, turn to a partner, and share. • SW then raise their white boards to show the teacher their completed equation. • TW continue to have students write and solve using white boards going from one-digit to two-digit examples.	Student solutions Student equations Student equations and responses in pairs Student problems and solutions

(continued)

SIOP® LESSON PLAN *(continued)*

Language Objectives:	Meaningful Activities:	Review/Assessment:
1. SW demonstrate analysis of adding and subtracting whole numbers by writing addition and subtraction sentences and orally sharing with a partner.	• TW remind students that we also use words to write and share equations. • TW go back to the book and explain how words are used to formulate the addition problems and then to solve. For example: *Problem*: If you have three M&M's and are given two more, how many M&M's do you have all together? *Solution*: Three M&M's plus two more equals five. • TW give two examples and have students solve on their white boards in small groups and share. • TW then list four problems on the overhead, two addition and two subtraction. • SW write the appropriate solution on one side of the sentence strip and the equation on the other. • SW present one strip to a partner.	Students' sentence strips and responses Student presentations

Wrap-up: Ticket out:
SW formulate one problem, addition or subtraction, and its solution.

Another great interaction technique is "Find Your Match." Notice the level of engagement for students that the technique provides.

Find Your Match (adapted from *99 Ideas,* page 112)

SIOP® COMPONENT: Interaction

Grade Level: All
Subject Levels: All
Grouping Configurations: Partners, small groups, whole class
Approximate Time Involved: 15–30 minutes, depending on how it is used
Materials: An index card for each student

Description

"Find Your Match" encourages interaction among class members as they read and produce oral language, practice newly learned content, or review for a test. Each student is given a card with information that matches the information on another student's card.

Grade 9–12 Math Activity Application

Lesson Concept: Representing Linear, Quadratic, and Exponential Functions in Multiple Forms

Content Objectives:
SW represent linear, quadratic, and exponential functions in multiple representations.

Language Objectives:
SW orally state what type of function is represented by three matching cards.

Lesson Sequence:

1. On thirty index cards, represent 10 functions (linear, quadratic, exponential) algebraically, graphically, and tabularly. Provide each student with a card.

2. Have students mix with each other, reading aloud the information on their cards.

3. Once students have had several opportunities to share their information, call time and instruct students to find their matches.

4. When students with a match find each other, have them move to the side of the room and determine whether they have a true match and what type of function their matching cards represents.

5. To conclude the activity, have partners read their cards to the rest of the class and state whether their matches represent a linear, quadratic, or exponential function.

Activity Differentiated Based on Levels of Language Proficiency

Beginning:

- For this activity, provide students with pictures of linear, quadratic, and exponential functions. Students at the beginning stages of second language acquisition can point to the picture to express what type of function is represented by the group's matching cards.

Intermediate:

- English learners at the Intermediate levels also benefit from pictures of linear, quadratic, and exponential functions. Students at this level of language proficiency can point to the picture as they complete the following cloze sentence:

  The function is _____

Advanced High:

- This group of students can complete the following cloze sentence to determine what type of function the matching cards represent:

  The three matching cards represent a _____ function.

SIOP® Math Techniques and Activities: Practice & Application

One of the aspects of effective teaching that is often overlooked or underutilized is planning sufficient time for students to practice and apply new concepts, procedures, language and/or skills. Typically, teachers present material and then have students complete some type of worksheet containing a series of math problems. However, English learners (and often other students) require significant opportunities to practice and apply new material with teacher supervision. In this next section we present a Practice & Application activity and lesson plan.

"Bingo" is a time honored tradition in many classrooms. It is a fun game for students of all ages, but at the same time, it provides excellent practice with important math facts and vocabulary.

Bingo

COMPONENT: Practice & Application

Grade Level: All
Grouping Configurations: Individual
Approximate Time Involved: About 15 minutes
Materials: Piece of paper for each student; paper squares or cereal shaped in O's

Description

The purpose of "Bingo" is to provide students with hands-on practice with words or facts. Model for students how to fold a blank piece of newsprint into 9 (3 × 3) or 16 (4 × 4) squares and display 10–20 vocabulary words (or math facts). Students fill in the squares in random order so that no two are identical. The teacher passes out paper squares (which can be collected and saved for the next Bingo game) or loop cereal (which can be eaten after the lesson). While the game is in session, do not call out the exact word or fact the students have written, but a definition or related fact instead. Students have to find the match and cover the square on the Bingo sheet with the small paper square or cereal O. The first student to get three or four in row, column, or diagonal, calls out "Bingo" and explains why he or she won.

Grade 6–8 Math Activity Application

Lesson Concept: Formulas of Area, Surface Area, and Volume of Geometric Figures

Content Objectives:
SW identify the formulas for area, surface area, and volume of geometric figures.

Language Objectives:
SW write verbal descriptions of formulas representing area, surface area, and volume of geometric figures.

(continued)

Lesson Sequence:

1. On the board or overhead write 10 to 20 formulas representing the area, surface area, and volume of geometric figures. Have students work in pairs to determine the description of the formula. (Example: If the formula is $A = l \times w$, students write, "Area of a rectangle.") Collect students' information.

2. Provide each student with a copy of a 3×3 or 4×4 array and paper squares or loop cereal. Instruct students to fill in the squares in random order with the formulas written on the overhead. Inform students that not all formulas will be used and that each chosen formula can only be written once.

3. In a paper bag, place the 10 to 20 descriptions of the formulas. Randomly select a description from the bag and read it aloud to the class. If students have the matching formula on their array, instruct them to use a square paper or a piece of loop cereal to cover the formula.

4. This process continues until a student covers three or four in a row, column, or diagonal. At this time, the student calls out "Bingo" and explains how he/she won.

Activity Differentiated Based on Levels of Language Proficiency

Beginning:

- When students are working on writing the description of the formula, consider having them draw a picture instead of writing a description.

- During the "Bingo" activity, read the description and show a pictorial representation of the formula. For example, if the description reads "Area of a square," consider showing a picture of a square with the area colored in yellow.

- If a student wins, he or she can simply say, "Bingo." You can double check the work to validate the Bingo.

Intermediate:

- Students at the Intermediate level benefit from drawing and seeing pictures. Therefore, allow them to draw a picture of the formula and write a description of the formula using the cloze sentence:

  _____ of a _____ . (Fill in first blank with: Area, Surface Area, or Volume; Fill in second blank with: any polygon, three-dimensional figures, including cylinder and spheres.)

- If a student wins a Bingo, he or she can use the following sentence frame:

  Bingo! I win because. . . .

Advanced High:

- Students at this level may decide to use a sentence frame or simply state how they won Bingo:

  Bingo! I win because. . . .

SIOP® Math Techniques and Activities: Review & Assessment

On the SIOP® protocol, Lesson Delivery follows Practice & Application. The features of Lesson Delivery are subsumed in the lessons presented in Chapters 3 and 4, so that a separate section isn't necessary. The features of Lesson Delivery include: content and language objectives clearly supported by lesson delivery, students engaged approximately 90% to 100% of the lesson, and appropriate pacing. As you can see from these features, they are expected to be part of each of the lessons presented.

Following Lesson Delivery is the Review & Assessment component. Notice how this particular activity touches on the components of Practice & Application and Interaction in addition to Review & Assessment.

The "Find Someone Who" activity is a great way to involve students in assessing their own knowledge of the math concepts or processes from the lesson—and to find help from others.

Find Someone Who (adapted from *99 Ideas,* pg. 182)

SIOP® **COMPONENT:** Practice & Application, Interaction, and
Review & Assessment

Grade Levels: K–12
Grouping Configurations: Partners
Approximate Time Involved: About 10–15 minutes
Materials: Completed written task

Description

"Find Someone Who" gives students an opportunity to review material and clarify answers they don't know. The activity can be structured so that once students have found a partner, they can share a completed task or formulated responses and get feedback from their peers. As students are sharing, it also provides an opportunity for the teacher to assess students' understanding of language and content objectives.

Students are given a characteristic to use to match up with another student in the classroom. For example:
- Find someone who is wearing a similar color or style of shoe that you are.
- Find someone whose last name starts with the same letter as yours.
- Find someone who has the same type of pet that you have.

Once students are partnered up, they share their formulated responses, or completed academic task, always in a complete sentence. The teacher must remember to provide them with the frame they should use to share.

This activity may also be used as an opportunity for students to provide one another feedback. After one partner shares, the teacher might have the other respond by:
- Agreeing or disagreeing with something their partner said.
- Summarizing or paraphrasing what their partner said.
- Stating something they liked or learned.
- Posing a question about something that their partner shared.

Grade 3–5 Math Activity Application:

Lesson Concept: Analyzing and Describing Characteristics of Geometric Shapes

Content Objectives:

SW demonstrate comprehension of two- and three-dimensional geometric shapes by comparing and contrasting triangles based on their sides and angles.

SW demonstrate knowledge of two- and three-dimensional geometric shapes by recalling different geometrical shapes and telling how they are different and the same.

Language Objective:

SW demonstrate synthesis of two- and three-dimensional geometric shapes by creating geometric shapes using triangles and explaining in a "Find Someone Who" activity using the frame:

I created a _____ by . . .

In the lesson example (Analyzing and Describing Characteristics of Geometric Shapes), "Find Someone Who" is the meaningful activity in the Language Objective of the lesson. Students are provided a structure to demonstrate their content understanding and to practice using the language, including key vocabulary and process/functional vocabulary.

Activity Differentiated Based on Levels of Language Proficiency

Beginning:

- Give students a preview in their L1 of the academic language that they will need to understand for the lesson. Include key content vocabulary and process/functional vocabulary.
- Students have worked in pairs to create shapes, and then done so individually. When possible, give them the opportunity to work with a language broker (a student who speaks both the L1 of your ELs and English).
- When it comes time to share similarities and differences, modify their general frame to: These are _____ . (They plug in *similar* or *different*.)
- When it comes time to share their shapes, modify their specific frame. Limit it to: I created a _____. (They share the shape and not the process.)
- Also consider having students share in their L1 and English. Be sure to give them a frame for the L1 as well so that they are always expected to speak in complete sentences.
- The teacher should guide students in practicing the frames before they have to share. *Note: This should be done for all language levels and native English speakers.*

Intermediate:

- Give students a preview in their L1 of the academic language they will need to understand for the lesson. Include key content vocabulary and process/functional vocabulary.

- Have students share only one shape and the specific frame
- Also consider having them share in both their L1 and English. Be sure to give them a frame for the L1 as well, so that they know they are always expected to speak in complete sentences.

Advanced High:

- All of the modifications made for the Intermediate students would also be appropriate for this group.
- These students should be able to share and complete all frames with their partners if the modeling outlined throughout the lesson was consistent and ongoing.

In the following lesson, you will see "Find Someone Who" used effectively in a math class.

SIOP® LESSON PLAN

Key: SW = Students will; TW = Teacher will; HOTS = Higher Order Thinking Skills (questions and tasks)

Lesson: Analyzing and Describing Characteristics of Geometric Shapes
Grades: 3–5

NCTM Standard: Analyze characteristics and properties of two- and three-dimensional geometric shapes and develop mathematical arguments about geometric relationships.

Expectation:

1. Identify, compare, and analyze attributes of two- and three-dimensional shapes.
2. Build and draw geometric objects.

Visuals & Resources:

Overhead, pattern blocks, ruler, protractor, BLM 4, glue, and construction paper

Key Vocabulary:	General Frames:
properties, square, triangle, rhombus, trapezoid, hexagon, characteristic, similar, different	They are similar/different because . . .
	Specific Frames:
HOTS (Higher order thinking questions or tasks): What is something you learned about triangles today?	I created a _____ by . . .

Connections to Prior Knowledge/ Provide Background Information:
TW ask students to think about what they know about shapes. SW share with a partner at their table and then share out with the whole group.

Content Objectives:	Meaningful Activities:	Review/Assessment:
1. SW demonstrate comprehension of two- and three-dimensional geometric shapes by comparing and contrasting triangles based on their sides and angles.	• TW display various triangular shapes and ask questions, for example: What are these shapes? How do you know? • SW answer questions in small groups, then share out with the whole group. Responses should include:	Student responses

2. SW demonstrate knowledge of two- and three-dimensional geometric shapes by recalling different geometrical shapes and telling how they are different and the same.	They have three sides, three corners and angles. The sides are straight, etc. • TW distribute triangular figures in different sizes and ask students in their small groups to talk about how they are the same and how they are different. • TW remind them to use their "tools" to measure sides and angles to identify similarities and differences. • SW share in groups and come to a consensus on a response using the frame: They are similar/different because . . . • TW remind students of the other geometric shapes that they have studied, reviewing key vocabulary from above. • TW distribute pattern blocks and talk about what the shapes have in common and how they are different. • SW share in groups and come to consensus on a response using the frame: They are similar/different because . . .	 Student responses and appropriate use of the general frame Student responses and appropriate use of the general frame

Language Objectives:

1. SW demonstrate synthesis of two- and three-dimensional geometric shapes by creating geometric shapes using triangles and explaining in a "Find Someone Who" activity using the frame: I created a _____ by . . .	• TW give students BLM 4 and ask them to cut out the triangle shapes and create any two of the geometric shapes they have studied. • SW cut out their triangles (BLM 4) and create two geometric shapes with their triangles and glue them on construction paper. • SW then be asked to "Find Someone Who" is wearing a similar style and color of shoe to partner up with them. • SW share their shapes using the specific frame: I created a _____ by . . . • SW then be asked to "Find Someone Who" is wearing a similar style of shirt to partner up with them. • SW again share their shapes using the specific frame: I created a _____ by . . . • *Note: Have students find "someone who" several times. Students will be practicing key vocabulary as they complete their frames.*	 Student created shapes Student shapes and appropriate use of specific frame.

Wrap-up: TW have students answer the HOTS question, "What is something you learned about triangles today?" SW hand it in on their way out the door.

Concluding Thoughts

In Chapters 3 and 4 we presented proven activities and techniques for SIOP® lessons. As we emphasized previously—and it bears repeating—simply adding new techniques and activities to a lesson most likely won't give students the kind of cohesive, effective instruction they need to make progress academically. All components of the SIOP® Model and its thirty features work together to make a lesson effective for English learners. Although we have not identified and isolated each and every feature in the math lesson plans in this chapter, our desire is that educators can envision how these features have been thought through in the planning process and would be apparent in the delivery of the lesson.

We hope you will put to use the range of techniques presented in these chapters that will advance your students' academic English skills in math and make the curriculum topics of this specialized area comprehensible for all students.

chapter 5

Lesson and Unit Design for SIOP® Mathematics Lessons

Grades K–2

Melissa Castillo

Math Unit, Grades K–2

Overview of the Unit

As young learners are introduced to new math concepts, it is important that the instruction be geared to their unique learning needs. You will note that in each of the lessons, students are introduced to very abstract mathematical concepts by means of a very concrete approach. Each lesson also includes what might be considered challenging academic vocabulary and language frames, especially for young students. However, research consistently shows that in order for students to achieve, they must have a rich and varied vocabulary. Certainly, the language included in the lessons is complex; however, if it is explicitly taught and practiced, students can acquire an understanding of the concepts and develop academic language.

The following unit has four lessons taught across five days. In the first lesson, there is an emphasis on key content vocabulary. As students move through the week's lessons, that particular set of vocabulary words is reinforced, and expanded upon to support the content of each lesson. For example, in Lesson 1, the key vocabulary included the terms *measure, measurement, nonstandard, standard,* and *unit* and in Lesson 2, those same terms are repeated as key vocabulary, but we added *length, foot, feet,* and *inches.*

Several of the lessons also include a literature connection. English learners benefit from exposure to cross-curricular connections throughout the school day, and math is an area that lends itself to the inclusion of literature. Coupling literature with math also provides the opportunity for students to practice language skills in all domains: reading, writing, listening, and speaking.

This unit of study is guided by the questions: What are nonstandard and standard units of measurement? How are they used for measurement (i.e., length, distance, and area)? What is the academic vocabulary needed for students to articulate their understanding of nonstandard and standard units of measurement? By the end of the unit, students will have developed the knowledge and academic language needed to address these questions.

Throughout the unit, you will also notice that there are Planning Points and Think-Alouds. As experienced SIOP® teachers, we have included in the lessons our own thought processes for planning SIOP® lessons that will result in the high-quality teaching described above. The purpose of the Planning Points is to highlight the process of planning a SIOP® unit and to clarify any questions you may have about each specific lesson plan. The Think-Alouds are questions we asked ourselves or issues that arose throughout the SIOP® unit planning process. We thought it would be helpful to include them as a possible resource for your own planning.

We hope that you will use the models we have provided in this chapter as a guide for your own planning and teaching. You may find that some of the lessons will require more or less time than what is allotted here, depending on your students. That is fine; the lessons are designed to show how SIOP® features are integrated into math lessons over the course of a unit of study. We encourage you to adapt them for your own students' needs.

SIOP® Planning Flow Chart

Grade Level: 1
Unit Concept (Big Idea): Using Nonstandard and Standard Units of Measurement

Subject: Math
Approximate Time Involved: 5 days 50–60 minute math lessons

Lesson Focus Day 1:
Develop content-specific vocabulary to articulate how and why to use nonstandard and standard units of measurement.

Content Objective:
SW demonstrate knowledge of key content vocabulary by defining what the words *measurement, measure, unit, nonstandard,* and *standard* mean in their own words.

Language Objective:
SW demonstrate application of key content vocabulary by identifying orally and in writing examples of nonstandard and standard units using the frame:
An example of a nonstandard/standard unit of measurement is   .

Reading/Writing/Discussion Activities:
Turn to Your Partner/Conga Line

Lesson Focus Days 2 & 3:
Using nonstandard and standard units of measurement

Content Objective:
SW demonstrate comprehension of using nonstandard and standard units of measurement by comparing measurements of nonstandard (hands and feet) and standard (ruler) units on a graphic organizer.

Language Objective:
SW demonstrate evaluation of using nonstandard and standard units of measurement by explaining which unit of measurement is most accurate and orally sharing their measurements in a "You Are the Teacher" activity.

Reading/Writing/Discussion Activities:
You Are the Teacher, Turn to your Partner

Lesson Focus Day 4:
Using nonstandard and standard units of measurement to determine length and distance

Content Objective:
SW demonstrate comprehension of the attributes of length and distance by measuring the length and distance between various classroom objects.

Language Objective:
SW demonstrate analysis of the attributes of length and distance by comparing their measurements orally and in writing using the words *shorter, longer, taller,* and *wider.*

Reading/Writing/Discussion Activities:
Small Group/Tickets Out

Lesson Focus Day 5:
Using nonstandard and standard units of measurement to measure area

Content Objective:
SW demonstrate comprehension of the attributes of area by outlining shapes on a poster board and predicting how many nonstandard units equal its area.

Language Objective:
SW demonstrate analysis of the attributes of area by concluding whether their predictions were accurate and orally sharing using the frames:
I predicted the area of my shape would measure _____ units.
My shape measured _____ units.
The area is _____ units (cotton balls).

Reading/Writing/Discussion Activities:
Specific Frames/Think, Pair, Share
Please note that the BLMs for this unit are in Appendix C.

SIOP® LESSON PLAN, *Grade 1, Day 1*

Key: SW = Students will; TW = Teacher will; HOTS = Higher Order Thinking Skills (questions and tasks)

Unit Focus: Using Nonstandard and Standard Units of Measurement

Lesson 1: Develop Content-Specific Vocabulary to Articulate How and Why to Use Nonstandard and Standard Units of Measurement
Grade: 1

NCTM Standard: Understand measurable attributes of objects and the units, systems, and processes of measurement.

Expectation:
Understand how to measure using nonstandard and standard units.

(continued)

SIOP® LESSON PLAN, *Grade 1, Day 1* (continued)

Visuals & Resources:

"Inch by Inch" by Leo Lionni (book)
BLM 5, Vocabulary Activity Sheet

Brown paper bags containing a variety of standard and nonstandard units of measure.

Key Vocabulary:	*General Frames:*
measure/measurement, nonstandard, standard, unit	_____ means _____.
HOTS (Higher order thinking questions or tasks):	**Specific Frames:**
What have you measured? Why is it important to be able to determine the measurement of something?	An example of a nonstandard/standard unit of measurement is _____.

Connections to Prior Knowledge/ Provide Background Information:
TW read the book "Inch by Inch" to introduce the unit. TW explain that when we measure things around us, we can use either type of measurement. TW ask students to think about things they have measured in school or outside of school and what they used to measure. SW turn to a partner and share using the frame: I have measured a _____. I used a _____ to measure it.

Content Objectives:	*Meaningful Activities: Lesson Sequence*	*Review/Assessment:*
1. SW demonstrate knowledge of key content vocabulary by defining what the words *measurement, nonstandard* and *standard* mean in their own words.	• TW remind students of the book they read and how the inchworm was used to measure different things. • TW explain that there are certain words that need to be understood and defined to share how to measure and what we use to measure. • TW explicitly introduce the words *nonstandard, standard, unit,* and *measurement.* • TW model on the overhead examples of each (for example, cubes, string, cut-out worm from the story, pencils, ruler, yardstick, etc.) • SW work on the Vocabulary Activity Sheet (BLM 5), using their own words to define each key vocabulary word, illustrate an example or examples, and complete the listed frame: _____ means _____.	Student vocabulary sheets

Language Objectives:		
1. SW demonstrate application of key content vocabulary by identifying orally and in writing examples of nonstandard and standard units using the frame: An example of a nonstandard/standard unit of measurement is _____.	• TW place a brown bag in the middle of each table and ask students to take out the items one by one and talk about what type of unit of measurement it is an example of. • SW use the specific frame to share at their tables and to complete in writing their Vocabulary Activity Sheet. • SW share their activity sheet in a Conga Line.	Teacher observations of group responses Student responses and appropriate use of frames Teacher observation of student sharing in Conga line

Wrap-up: Rate your learning: TW ask students to think about how well they've met the learning objectives. SW show one finger if they haven't learned or met the objectives, two fingers if they have met the objectives.

SIOP® Features:

Preparation
X Adaptation of content
X Links to background
X Links to past learning
X Strategies incorporated

Integration of Processes
X Reading
X Writing
X Speaking
X Listening

Scaffolding
X Modeling
X Guided practice
X Independent practice
X Comprehensible input

Application
X Hands-on
X Meaningful
X Linked to objectives
X Promotes engagement

Grouping Options
X Whole class
__ Small groups
X Partners
X Independent

Assessment
X Individual
X Group
X Written
X Oral

PLANNING POINTS for SIOP® Lesson Plan, Grade 1, Day 1

- Objectives should be introduced and reviewed with students before, during, and at the conclusion of the lesson. (Refer to Figure 2.3 in Chapter 2 for ideas on how to make objectives a relevant part of lessons.) As we know from Bloom's research and subsequent taxonomy, students need to experience a variety of levels of thinking, not just the lowest level—simple recall of information. You will note in the objectives that we provide opportunities for students to use thinking from all levels of Bloom's Taxonomy. (See Chapter 6, page 88 for a list of Bloom's levels.)

- In order to determine what language objective is appropriate, I considered the needs of all students based on their language and literacy levels. The language objective in this lesson allows for modification and adaptation for all learners. For example, students with limited language might only use the frame orally and not in writing, whereas students with more proficiency do both. What is important is that through the practice of language, students are given the opportunity to demonstrate both their understanding of the content as well as their growth in language development.

- Modification of frames must be considered throughout the unit.

- Conga Line is an activity used to promote the practice of the key concepts in a lesson—in this case key vocabulary—while promoting oral language development. Students form two parallel lines facing one another. One line is identified as group A and the other as group B. Group A shares whatever responses they've been asked to formulate, while group B listens; then group B shares while group A listens. When both have shared, the student at the end of the line in group A "congas" down the center of the parallel lines while everyone else in his or her line takes a step to the left to even out the lines one more time. Students end up with a new partner facing them and the process begins again. Have students repeat the process of changing partners several times. I use this activity because it is a fun way to maximize language practice and involve all students in the lesson.

THINK-ALOUDS for SIOP® Lesson Plan, Grade 1, Day 1

- How much practice will students need with the initial vocabulary in order to effectively use it in this lesson as well as in other lessons in the unit?

- How much explicit instruction and practice of the frame will need to be provided? And, how will I modify the frame for the different levels of language learners in the classroom? For example: *Students at intermediate and proficient levels can complete:* "An example of a nonstandard/standard unit of measurement is _____." *Beginners can complete*: "A nonstandard/standard unit is_____."

- How will I modify or adapt the Vocabulary Activity Sheet for different language levels? I think my students at intermediate and proficient levels can complete the activity sheet as is. Students at beginning levels of English proficiency will copy the teacher definition and will illustrate the word.

SIOP® LESSON PLAN, *Grade 1, Days 2–3*

Key: SW = Students will; TW = Teacher will; HOTS = Higher Order Thinking Skills (questions and tasks)

Unit Focus: Using Nonstandard and Standard Units of Measurement (Reference Chapter 3, K–2 Using Nonstandard and Standard Units of Measurement)

Lessons 2 & 3: Using Nonstandard and Standard Units of Measurement
Grade: 1

NCTM Standard: Understand measurable attributes of objects and the units, systems, and processes of measurement.

Expectations:
1. Understand how to measure using nonstandard and standard units.
2. Recognize length as a measurable attribute of objects.

Visuals & Resources:
"How Big Is a Foot?" by Rolf Miller (book)
BLM 6 Let's Measure the Length Of?
Rulers
"You Are the Teacher" *99 Ideas and Activities for Teaching English Learners with the SIOP® Model*

Key Vocabulary:	General Frames:
nonstandard, standard, measurement, unit, length, foot, feet, inches	I can use . . .
HOTS (Higher order thinking questions or tasks): Which unit of measurement provides the most accurate measurement? Why?	*Specific Frames:* The best way to measure is using a _____ because_____.

Connections to Prior Knowledge/ Provide Background Information:
TW introduce new content vocabulary and have students take some time to complete the Vocabulary Activity Sheet with new terms. TW ask students if they have ever wanted to know how long, tall, or wide something was. How did you figure it out? Why was it important? SW turn to a partner/s and share. Read the Story "How Big Is a Foot?"

Content Objectives:	Meaningful Activities: Lesson Sequence:	Review/Assessment:
1. SW demonstrate comprehension of using nonstandard and standard units of measurement by comparing measurements of nonstandard (hands and feet) and standard (ruler) units on a graphic organizer.	• TW read the book "How Big Is a Foot?" • TW ask questions throughout the story about why the apprentice in the book is unable to make a bed that fits the queen. • In small groups, SW answer and share out. • TW ask students what they might have to use for measurement besides feet, as the apprentice did in the story. • SW answer in groups using the general frame: I can use. . . . (Teacher must explicitly teach the frame, model, and guide students in practice.) • TW then explain to students that they have tools that they can use to measure such as their own hands, feet, and fingers. • TW also explicitly re-teach and emphasize the vocabulary term *nonstandard*, making sure students understand that these tools are nonstandard units used to measure. • TW explain that they will be using these tools to measure and model several examples. • In small groups, SW use their nonstandard units (hands, feet and fingers) to measure items such as a desk, length of the chalkboard, and a book. Note: Since students are working in groups, ask each of them to put their name next to their measurement. • SW list their measurement on chart paper using the format of the graphic organizer (BLM 6). • TW bring students back together and ask them if there are any other tools that they can use or that the apprentice could have used that aren't part of them. • SW respond in small groups and share out using the general frame: I can use . . . • TW share with them or remind them, depending on the grade level, about the ruler (standard unit) and how we can use it to measure in inches and feet. • TW also explicitly re-teach and emphasize the vocabulary words *standard* and *unit*. • TW model measuring with the ruler the same examples she used for nonstandard. • SW then be asked to measure with the ruler the same objects as in the earlier example. • In their groups, SW now add to their graph on the chart paper what the new measurements are using the ruler.	Student responses Teacher observation of student responses Student responses and appropriate use of the frame Measurements on chart paper and completed graphic organizers Student responses and appropriate use of the general frame Student measurements on graphic organizer

(continued)

SIOP® LESSON PLAN, *Grade 1, Days 2–3* (continued)

Language Objectives:	Meaningful Activities: Lesson Sequence:	Review/Assessment:
1. SW demonstrate evaluation of using nonstandard and standard units of measurement by explaining which unit of measurement is most accurate and orally sharing their measurements in a "You Are the Teacher" activity.	• TW ask students if they notice a difference in the measurement results that they have listed. • SW respond in groups and share out. • TW also ask them if there is one way to measure that might be better than another, and why? • SW answer in their groups and share out. • Students should come to the conclusion that nonstandard units of measurement are less reliable than standard units of measurement. • In their groups, SW complete the following specific frame at the bottom of their chart: The best way to measure is using a _____ because . . . . • In their groups, SW then set up their completed graphs at a designated station in the room. For example, if there are four groups, there will be four stations, one in each corner • SW leave one of their peers at their station and rotate to the nearest station clockwise. Note: The teacher can have students letter off or count off and randomly select students, students can choose within their groups, or teacher can designate who will stay each rotation. • The student who stays at the station will present the poster and explain which measurement was most accurate and why. SW continue to rotate until they return to their original station.	Student responses of their measurements (Their initial understanding of why one way to measure is more accurate than the other.) Student responses and appropriate use of the specific frame Student presentations (How they convey their work and understanding of it)

Wrap-up: TW ask students to answer the following questions on a note card:
What is an example of a nonstandard unit of measurement?
What is an example of a standard unit of measurement?
SW use the frames: A nonstandard example is _____. A standard example is _____.

SIOP Features:

Preparation
X Adaptation of content
X Links to background
X Links to past learning
X Strategies incorporated

Scaffolding
X Modeling
X Guided practice
X Independent practice
X Comprehensible input

Grouping Options
X Whole class
X Small groups
X Partners
X Independent

Integration of Processes
X Reading
X Writing
X Speaking
X Listening

Application
X Hands-on
X Meaningful
X Linked to objectives
X Promotes engagement

Assessment
X Individual
X Group
X Written
X Oral

PLANNING POINT for SIOP® Lesson Plan, Grade 1, Days 2–3

- "You Are the Teacher" activity in the language objective is introduced and explained in Chapter 4, page 47.

THINK-ALOUDS for SIOP® Lesson Plan, Grade 1, Days 2–3

- Since this is a two-day lesson, the content and language objectives stay the same for both days. The language objective will begin in Day 1 and carry over to Day 2. What should I consider in determining the language objective for these lessons? One consideration is: How much language practice will it take for students to be able to articulate their understanding using academic vocabulary? Another consideration is: What kinds of behaviors will I observe in order to determine the students' level of understanding? For example: Do they understand the concept well enough to move forward? How will I know this? I think if they meet the language objective, I'll get the information I need.

- How am I building on the initial content vocabulary to make sure that students continue to practice and learn the new vocabulary necessary for this lesson?

- How will I determine whether students have the background experiences to answer my question in the Providing Background section of the lesson plan? If they don't, will enough background be provided by their peers when they share with each other?

- How will I group my students to ensure that they will all successfully move through the "You Are the Teacher" activity, regardless of their language levels?

- Will the connection between the story used to introduce the concept and the concept be clear to students? Have I planned for that explicit connection? How will I assess that it has been made?

- Have I provided ample opportunity for students to practice the concept of standard and nonstandard units of measurement so that they can successfully articulate their understanding of it in order to meet the Language Objective?

- Will this lesson need to be carried over an additional day so that students will be able to demonstrate their learning to the level necessary to move on to Lessons 4 and 5?

- How will I assess students' learning throughout the lesson?

SIOP® LESSON PLAN, *Grade 1, Day 4*

Key: SW = Students will; TW = Teacher will; HOTS = Higher Order Thinking Skills (questions and tasks)

Unit Focus: Using Nonstandard and Standard Units of Measurement

Lesson 4: Using Nonstandard and Standard Units of Measurement to Determine Length and Distance
Grade: 1

NCTM Standard: Understand measurable attributes of objects and the units, systems, and processes of measurement.

Expectations:
1. Understand how to measure using nonstandard and standard units.
2. Recognize the attributes of length, distance, volume, weight, area, and time.

SIOP® LESSON PLAN, *Grade 1, Day 4* (continued)

Visuals & Resources:
BLM 7 Measuring Length and Distance

Key Vocabulary:	*General Frames:*
nonstandard, standard, measurement, unit, length, foot, feet, inches, distance	One thing I remember is . . .
HOTS (Higher order thinking questions or tasks): Comparing the distance of classroom objects.	**Specific Frames:** The distance between _____ and _____ is . . . . _____ is longer than _____. _____ is shorter than _____. _____ is taller than _____. The distance between _____ and _____ is further than. . . .

Connections to Prior Knowledge/ Provide Background Information:
TW ask students to think back to the book "How Big Is a Foot?" read in the previous lesson and talk at their tables about what the king's problem was and how he solved it. SW share in their small groups and then share out to the whole class. TW also introduce the new key content word (distance) and have students complete the Vocabulary Activity Sheet from Lesson #1.

Content Objectives:	*Meaningful Activities: Lesson Sequence:*	*Review/Assessment:*
1. SW demonstrate comprehension of the attributes of length and distance by measuring the length and distance between various classroom objects and listing them on a table (graph).	• TW remind students that in the previous lesson they measured the length of different items in the classroom using standard and nonstandard units. • TW explicitly refer back to the charts the students had completed in the previous lesson. • SW take two minutes to turn to a partner and share one thing they remember from yesterday's lesson using the frame: One thing I remember is . . . • TW explain that they will continue to measure classroom items using standard and nonstandard units, but that the focus is not only the length of each item but the *distance* between different items in the class. • TW remind students of what is meant by the length of an object. • TW introduce what is meant by the distance between two or more objects. • TW also give explicit examples and model how to measure using both types of measurement units (standard and nonstandard). • SW be given tools for standard and nonstandard units of measurement to choose from (resources from previous lessons). • SW complete the measuring length and distance table in pairs (see BLM 7). Note: Even though they are working in pairs, each student should fill in his or her individual table to use for the language objective.	Teacher observation and student responses Student use of the charts (references) to share what was learned Student table and measurements listed, including length and distance, type of measurement, and accurate units

Language Objectives:

1. SW demonstrate analysis of the attributes of length and distance by orally and in writing comparing their measurements using the words *shorter, longer, taller, wider,* and *further.*

- TW introduce the process/functional vocabulary students will be using to share their comparisons of the items and their difference in length as well as distance. Words include *shorter, longer, taller, wider,* and *further.*
- TW go back to the examples used to introduce and distinguish between length and distance to model the vocabulary necessary to articulate which items are shorter, longer, etc.
- TW also explain to students that even though they did not measure the width of the classroom items, they can determine which is wider just by looking at them.
- TW also explicitly model the specific frame students will use to articulate their comparisons.
- SW use the measurements on their activity sheet to complete the following frames on the back of their Activity Sheet:

The distance between _____ and _____ is . . .
_____ is longer than _____.
_____ is shorter than _____.
The distance between _____ and _____ is further than. . . .

- SW share at least two frames with their table group.

Student responses and comparisons between what they have measured

Student application of the process/functional vocabulary

Student frames, how many can they effectively complete

Wrap-up: Ticket out: SW complete the following frames on a index card:
One thing I learned is . . .
Something I remembered from yesterday was . . .
I can use . . .

SIOP® Features:

Preparation
X Adaptation of content
X Links to background
X Links to past learning
X Strategies incorporated

Scaffolding
X Modeling
X Guided practice
X Independent practice
X Comprehensible input

Grouping Options
X Whole class
X Small groups
X Partners
X Independent

Integration of Processes
X Reading
X Writing
X Speaking
X Listening

Application
X Hands-on
X Meaningful
X Linked to objectives
X Promotes engagement

Assessment
X Individual
X Group
X Written
X Oral

THINK-ALOUDS for SIOP® Lesson Plan, Grade 1, Day 4

- How am I building on the initial content vocabulary to make sure that students continue to practice it and also learn the new vocabulary necessary for this lesson?

- Because this lesson directly builds on the previous lesson, how will I make sure that explicit connections are made to past learning?

- How will I pair students to ensure they are able to complete the academic task? How much language will be required? Will students at all the language levels in my class be able to participate fully?

- How will I modify the frames so that students from all language levels can complete them both orally and in writing?

Proficient and Intermediate Students will complete the frames:

The distance between _____ and _____ is . . . .
_____ is longer than _____.
_____ is shorter than _____.
_____ is taller than _____.

Beginners will complete:

The distance is _____ units.
_____ is longer.

- How many frames will students need to complete to demonstrate their mastery of the Language Objective? How will I assess them both independently and in their groups?

SIOP® LESSON PLAN, *Grade 1, Day 5*

Key: SW = Students will; TW = Teacher will; HOTS = Higher Order Thinking Skills (questions and tasks)

Unit Focus: Using Nonstandard and Standard Units of Measurement

Lesson 5: Using Nonstandard and Standard Units of Measurement to Measure Area
Grade: 1

NCTM Standard: Understand measurable attributes of objects and the units, systems, and processes of measurement.

Expectations:
1. Understand how to measure using nonstandard and standard units.
2. Recognize the attributes of length, distance, volume, weight, area, and time.

Visuals & Resources:
"Little Cloud" by Eric Carle (book)
Blue pieces of poster board cut in halves (one half for each small group of four students)
Cotton balls
Glue sticks
Wikki Wax Stixs (4 for each student) or any other resource (yarn, toothpicks, etc.) to outline shapes

Key Vocabulary:	General Frames:
nonstandard, standard, unit, measurement, length, foot, feet, inches, distance, area	I predict . . .

HOTS (Higher order thinking questions or tasks):	*Specific Frames:*
What other objects could we use to figure out the area of the shapes? Would the area of the shape be the same? Why or Why not?	I predicted my shape would measure _____ units. My shape measured _____ units. The area is _____ units.

Connections to Prior Knowledge/ Provide Background Information:

TW introduce new content vocabulary (*area*) and have students take some time to complete the Vocabulary Activity Sheet with the new term.

TW ask students to think about how they might figure out the area of something using nonstandard units. TW remind them to reference past vocabulary sheets and charts posted from previous lessons.

TW ask students if they think they can use any of the nonstandard units that they have used to measure previously. SW share their responses in small groups and then share out to the class.

Note: The goal is for students to begin to identify that the nonstandard units used up until now won't work and that they begin to consider what might work.

Content Objectives:	Meaningful Activities: Lesson Sequence:	Review/Assessment:
1. SW demonstrate comprehension of the attributes of area by outlining shapes on a poster board and predicting how many nonstandard units equal its area.	• TW read the book "Little Cloud" to the class. • TW elaborate on the different shapes that are formed by a cloud throughout the story. • TW explain that students will be creating shapes of their own (like in the book) and using a nonstandard unit of measurement to determine their shape's *area*. • TW explicitly introduce the concept of measuring *area*, referring back to the new content vocabulary word that was introduced at the beginning of the lesson. • TW explain that the nonstandard unit they will use to measure the area of their shapes is cotton balls, referring back to the clouds and how the shapes looked like they were full of cotton. • TW also explain that they are going to create shapes, predict how many nonstandard units it will measure (how many cotton balls it will take to fill the shape), fill the shape with the cotton balls, and confirm whether their predictions were accurate. • TW model how to use the Wikki Stixs to outline different shapes. • TW model how to make a prediction for the shape using the frame: I predict . . . (General frame allows for the teacher to decide how the prediction should be completed based on his or her students.) • TW fill the shape with cotton balls and have students count with her/him how many there are. • TW then explain that the number of cotton balls equals the area of the shape in nonstandard units.	

(continued)

SIOP® LESSON PLAN, *Grade 1, Day 5* (continued)

Content Objectives:	Meaningful Activities: Lesson Sequence:	Review/Assessment:
	• In small groups, SW create two shapes with their sticks on their poster board and label each shape. Note: Wax sticks should stick to the poster board; if not, have students glue into place.	
	• SW each predict how many cotton balls it will take to fill each shape on a sticky note using the frame:	Student responses and appropriate use of the frame
	I predict . . .	Student predictions
	• SW take turns orally sharing their predictions with their group. The strategy used for this prediction process is called Think, Write, Share. Note: Teacher always explicitly teaches the frame and models.	

Language Objectives:		
1. SW demonstrate analysis of the attributes of area by concluding whether their predictions were accurate and orally sharing using the frames: I predicted the area of my shape would measure _____ units. My shape measured _____ units. The area is _____ units (cotton balls).	• In their groups, SW glue the cotton balls into place. • SW count the number of cotton balls and label how many filled the shape below its name. • TW remind students that the cotton balls are being used as a nonstandard unit of measurement. • On another sticky note, SW each complete the specific frames, concluding whether their predictions were accurate. • SW post their sticky notes next to one another below the different shapes (their original prediction next to their conclusions). • SW take turns presenting their shapes, original predictions, and conclusions to the whole group.	Student-created shapes and labels (Are they able to identify the shape and number of cotton balls—nonstandard units?) Student responses and conclusions using frames appropriately Student presentations

Wrap-up: Students will answer the HOTS questions in a Think, Pair, Share.
What other objects could we use to figure out the area of the shapes?
Would the area of the shape be the same? Why or why not?

SIOP® Features:

Preparation
X Adaptation of content
X Links to background
X Links to past learning
X Strategies incorporated

Integration of Processes
X Reading
X Writing
X Speaking
X Listening

Scaffolding
X Modeling
X Guided practice
X Independent practice
X Comprehensible input

Application
X Hands-on
X Meaningful
X Linked to objectives
X Promotes engagement

Grouping Options
X Whole class
X Small groups
X Partners
X Independent

Assessment
X Individual
X Group
X Written
X Oral

PLANNING POINTS for SIOP® Lesson Plan, Grade 1, Day 5

- Area is a difficult concept for young students to grasp. The goal of this lesson is just to introduce the concept of area and give students a visual (concrete) representation of it.

- Because this is the last lesson in the unit, some type of formal assessment should be provided to wrap up the unit.

- This unit is only an introduction to measuring using standard and nonstandard units. This particular math standard will be the focus of instruction throughout the school year. It is also a concept that will continue to be developed as students move through second grade. The next unit would include using standard and nonstandard units to measure volume, weight, and time, expanding on the National Expectation from the National Standards on Mathematics: "Recognize the attributes of length, distance, volume, weight, area, and time."

THINK-ALOUDS for SIOP® Lesson Plan, Grade 1, Day 5

- How am I building on the initial content vocabulary to make sure that students continue to practice it and learn the new vocabulary necessary for this lesson?

- How will I make sure that students connect all of the dots? Did they make the connection between standard and nonstandard units, and using them to determine attributes (measurements) of length, distance, and area?

- How will I assess the development of the key content vocabulary that was introduced in Lesson 1 and emphasized in Lessons 2–5? Did students acquire enough of the lessons' vocabulary to understand the content concepts of the unit?

Concluding Thoughts

Teaching young children is exciting because, as one of their first teachers, you have an influence on their first school experiences, experiences that can have a long-term effect on their attitude toward school and learning. Offering students well-prepared, engaging lessons contributes to positive school experiences.

As you read through the unit, we hope that you noticed how abstract concepts were presented in concrete ways. The goal is to create learning situations where students are successful in acquiring the math skills and concepts required at their grade level. In meeting the goal, English learners will get off to a good start in building a solid foundation in mathematics.

You may find the level of detail in the SIOP® lesson plans overwhelming and think that you cannot possibly create such detailed plans every day for every subject area. You're right. The detail in these lessons was necessary to describe the exact steps in each lesson and to make the activities and processes comprehensible for readers. You undoubtedly will have your own shorthand and symbols to represent much of what was described in narrative form here. Regardless of the level of detail you use, it is helpful to keep the SIOP® protocol handy as you write lesson plans so that you make sure that all components are present in your lessons. And, as mentioned previously, the more practice you have in developing SIOP® lessons and units, the easier and more natural the process becomes.

Lesson and Unit Design for SIOP® Mathematics Lessons

Grades 3–5

Melissa Castillo

Math Unit, Grades 3–5

Overview of the Unit

In your process of planning and delivering instruction for students in the upper elementary grade levels, it is important that you continue to consider the specific needs of all learners. You will notice that the emphasis on academic vocabulary continues throughout this SIOP® mathematics unit. As the concepts and academic language become more complex, it is even more critical for teachers to provide explicit instruction coupled with modeling and plenty of practice and application. Another key piece that has been included in Lessons 3 and 4 of this unit is the teaching of learner strategies, specifically metacogni-

tive strategies—the process of purposefully monitoring our own thinking (Echevarria, Vogt & Short, 2008). Research indicates that good readers and learners use a variety of cognitive and metacognitive strategies; English learners should be taught to use these strategies in specific ways to support their learning both in the classroom and in outside settings.

This unit also plans for students to demonstrate their learning using different levels of critical thinking. As mentioned previously, Bloom's Taxonomy provides guidance for including a variety of levels of cognitive tasks in lessons. Bloom's Taxonomy includes the following levels:

1. *Knowledge*: arrange, define, duplicate, label, list, memorize, name, order, recognize, relate, recall, repeat, reproduce, state.

2. *Comprehension*: classify, describe, discuss, explain, express, identify, indicate, locate, recognize, report, restate, review, select, translate.

3. *Application*: apply, choose, demonstrate, dramatize, employ, illustrate, interpret, operate, practice, schedule, sketch, solve, use, write.

4. *Analysis*: analyze, appraise, calculate, categorize, compare, contrast, criticize, differentiate, discriminate, distinguish, examine, experiment, question, test.

5. *Synthesis*: arrange, assemble, collect, compose, construct, create, design, develop, formulate, manage, organize, plan, prepare, propose, set up, write.

6. *Evaluation*: appraise, argue, assess, attach, choose, compare, defend, estimate, judge, predict, rate, select, support, value, evaluate.

Research shows that most of the time, students are asked to answer questions or participate in academic tasks at only the lower levels of Bloom's Taxonomy. Of the approximately 80,000 questions that teachers ask annually, 80 percent are at the literal or knowledge level (Gall, 1984; Watson & Young, 1986). We recommend that you post the taxonomy in your classroom as a reminder to develop students' higher levels of thinking. Also, we suggest that you teach students what is meant by each level and how it supports critical thinking.

In each of the lessons in this unit, students are asked to formulate responses to questions at different levels as well as to participate in tasks or activities that are academically challenging. This not only ensures that they develop a deeper understanding of the concept, but because their responses require a great deal of detail, it also makes certain that students are using much more academic language than in a "typical" lesson.

This unit of study is guided by the questions: What are the characteristics of geometric shapes (i.e., trapezoid, hexagon, triangle, etc.)? How are geometric shapes the same and different? and What academic vocabulary is used to identify and describe different geometric shapes?

By the end of the unit, students will have developed the knowledge about geometric shapes and academic language necessary to address all three questions.

In the unit, you will also notice that there are Planning Points and Think-Alouds. As experienced SIOP® teachers, we have included in the lessons our own thought processes for planning SIOP® lessons that will result in the high-quality teaching described above. The purpose of the Planning Points is to highlight the process of planning a SIOP® unit and to clarify any questions you may have about each specific lesson plan. The

Think-Alouds are questions we asked ourselves or issues that arose throughout the SIOP® unit planning process. We thought it would be helpful to include them as a possible resource for your own planning.

We hope that you will use the models we have provided in this chapter as a guide for your own planning and teaching. You may find that some of the lessons will require more or less time than what is allotted here, depending on your students. That is fine; the lessons are designed to show how SIOP® features are integrated into math lessons over the course of a unit of study. We encourage you to adapt them for your own students' needs.

SIOP® Planning Flow Chart

Grade Level: 3	**Subject:** Math
Unit Concept (Big Idea): Analysis of Geometric Shapes	**Approximate Time Involved:** 5 days 50–60 minute math lessons

Lesson Focus Day 1:
Develop content-specific vocabulary in order to describe characteristics of two- and three-dimensional shapes.

Content Objectives:
SW demonstrate knowledge of the characteristics of two- and three-dimensional geometric shapes by defining what a square, triangle, rhombus, trapezoid, and hexagon are on a 4-Corners Vocabulary Chart.

Language Objective:
SW demonstrate comprehension of the characteristics of two- and three-dimensional geometric shapes by writing and explaining the characteristics of a square, triangle, rhombus, trapezoid, and hexagon using the frame:
One characteristic of a _____ is _____ .

Reading/Writing/Discussion Activities:
4-Corners Vocabulary Chart/BLM 8;Specific Frames/Turn to a Partner

Lesson Focus Day 2:
Comparing geometric shapes.

Content Objectives:
SW demonstrate comprehension of two- and three-dimensional geometric shapes by comparing and contrasting triangles based on their sides and angles.

SW demonstrate knowledge of two- and three-dimensional geometric shapes by recalling different geometric shapes and telling how they are different and the same.

Language Objective:
SW demonstrate synthesis of two- and three-dimensional geometric shapes by creating geometric shapes using triangles and explaining their creation in a "Find Someone Who" activity using the frame:
I created a _____ by . . .

Reading/Writing/Discussion Activities:
Find Someone Who/Specific Frames/Respond to

Lesson Focus Days 3 & 4:
Describing and drawing characteristics of two- and three-dimensional shapes.

Content Objectives:
SW demonstrate comprehension of two- and three-dimensional geometric shapes by illustrating and describing shapes they recognize in the poem "Shapes."

Language Objective:
SW demonstrate application of the characteristics of two- and three-dimensional geometric shapes by writing a poem including different shapes (rhombus, trapezoid, and hexagon) and orally sharing in small groups.

(continued)

SIOP® *Planning Flow Chart* *(continued)*

Reading/Writing/Discussion Activities:
Think, Pair, Share/White boards/Tickets Out

Lesson Focus Day 5:
Analysis of geometric shapes and application of content-specific vocabulary.

Content Objectives:
SW demonstrate application of the characteristics of two- and three-dimensional geometric shapes by constructing different examples from a story using geometric shapes.

Language Objective:
SW demonstrate evaluation of the characteristics of two- and three-dimensional geometric shapes by comparing different shapes and explaining their characteristics using key vocabulary.

Reading/Writing/Discussion Activities:
Grafitti Write/Specific Frames/Mix to Music

SIOP LESSON PLAN, *Grade 3, Day 1*

Key: SW = Students will. . .; TW = Teacher will; HOTS = Higher Order Thinking Skills (questions and tasks)

Unit Focus: Analysis of Geometric Shapes (Reference Chapter 4, Grades 3–5 Describing Characteristics of Two- and Three-Dimensional Shapes)

Lesson 1: Develop Content-Specific Vocabulary in Order to Describe Characteristics of Two- and Three-Dimensional Shapes
Grade: 3

NCTM Standard: Analyze characteristics and properties of two- and three-dimensional geometric shapes and develop mathematical arguments about geometric relationships.

Expectation:
1. Identify, compare, and analyze attributes of two- and three-dimensional shapes and develop vocabulary to describe the attributes.

Visuals & Resources:

"The Shape of Things" by Dayle Ann Dobbs (book) or

"Shapes, Shapes, Shapes" by Tana Hoban (book)

BLM 8 4-Corners Vocabulary from *99 Ideas and Activities for Teaching English Learners with The SIOP® Model*

BLM 1 What Do You Know about Geometric Shapes?

Chart paper and markers, pattern blocks for each group, and overhead blocks

Key Vocabulary:	*General Frames:*
square, triangle, rhombus, trapezoid, hexagon, characteristic	I noticed . . . Something that is the same is . . . Something that is different is . . .
HOTS (Higher order thinking questions or tasks): Why is it important to know what shapes surround us?	**Specific Frames:** One characteristic of a _____ is _____ .

Connections to Prior Knowledge/ Provide Background Information:
TW read one of the books listed above or a book of his or her choice on shapes to introduce the topic. TW ask students: "Where else do you see shapes inside or outside of your classroom?" SW answer in small groups and then share with the whole class.

Content Objectives:	Meaningful Activities: Lesson Sequence:	Review/Assessment:
1. SW demonstrate knowledge of the characteristics of two- and three-dimensional geometric shapes by defining what a square, triangle, rhombus, trapezoid, and hexagon are on a 4-Corners Vocabulary Chart.	• TW remind students of the shapes that were identified in the story and the shapes they discussed in their groups in the Building Background component above. • TW distribute pattern block sets to each of the groups. The set consists of a square, triangle, rhombus, trapezoid, and hexagon. • TW ask students to take a few minutes in their groups to talk about the different shapes and answer the following questions: What do you notice? How are they the same? How are they different? • SW share in their groups and answer using the general frames: I noticed . . . Something that is the same/different is . . . (Please note, teacher must explicitly teach and model the frame.) • TW then display each shape on the overhead and name each shape, explicitly identifying them as geometric shapes. • In their groups, SW share their ideas of what they notice about each shape as the teacher displays and names each one. • TW explicitly teach the word *characteristic* and explain to students that what they are sharing are the characteristics of each of the shapes. • In their groups, SW then be assigned one shape for which they will complete a 4-Corners Vocabulary Chart, BLM 8. • SW divide their piece of chart paper into four squares. • SW create a chart including an illustration (representing the shape), a definition (describing the shape), a sentence (that includes the word), and the word itself. • SW present their completed chart to the group. Teacher should post charts on a math word wall in the room so that students can use them for reference.	 Student responses and appropriate use of frames Completed 4-Corners Vocabulary Chart Presentation of charts

Language Objectives:

1. SW demonstrate comprehension of the characteristics of two- and three-dimensional geometric shapes by writing and explaining the characteristics of a square, triangle, rhombus, trapezoid, and hexagon using the frame:	• TW provide students a sheet (BLM 1) that has each shape drawn and labeled. • TW model and guide students in practicing how to appropriately use the specific frame. • SW complete the specific frame individually below each shape:	

(continued)

SIOP LESSON PLAN, *Grade 3, Day 1* *(continued)*

Content Objectives:	Meaningful Activities: Lesson Sequence:	Review/Assessment:
One characteristic of a _____ is _____.	One characteristic of a _____ is _____. • TW remind students to reference the 4-Corners Vocabulary Charts that they completed to help them complete their frames. • In pairs, SW share their completed frames.	Student responses on the BLM and appropriate use of frames in writing and when orally sharing Student responses in pairs

Wrap-up: Rate your learning: TW ask students to think about how well they've learned the different shapes. SW show one finger if they haven't learned or met the objectives; two fingers if they feel okay about their learning, have showed progress toward meeting the objective, but would like more information on shapes; and three fingers if they feel they have learned everything or met the objectives on geometric shapes.

THINK-ALOUDS for SIOP® Lesson Plan, Grade 3, Day 1

• I know that key content vocabulary must be taught and understood so that students will more readily and easily grasp the concept. Since this lesson is the introduction to the unit, how will I continue to emphasize and reinforce the vocabulary taught in this lesson throughout the unit?

• What additional language will have to be taught throughout the lesson so that students can describe the characteristics of the geometric shapes using academic language?

• In determining the language objective, I had to consider what opportunities I can give students to use the key vocabulary listed in the content objective. I also reflected on whether the language objective could be an opportunity to enrich their understanding of geometric shapes. In this case, it is. The sentence frame in the language objective brings both the key vocabulary and the concepts together, allowing students to articulate their understanding.

PLANNING POINT for SIOP® Lesson Plan, Grade 3, Day 1

• Throughout the unit, literature is used to introduce and/or reinforce the concept being taught. It is also an opportunity for students to practice the language domain of listening for a specific learning purpose.

SIOP® LESSON PLAN, *Grade 3, Day 2*

Key: SW = Students will . . .; TW = Teacher will; HOTS = Higher Order Thinking Skills (questions and tasks)

Unit: Analysis of Geometric Shapes (Reference Chapter 4, Lesson 3–5 Analyzing and Describing Characteristics of Geometric Shapes)

Lesson 2: Comparing Geometric Shapes
Grade: 3

NCTM Standard: Analyze characteristics and properties of two- and three-dimensional geometric shapes and develop mathematical arguments about geometric relationships.

Expectations:
1. Identify, compare, and analyze attributes of two- and three-dimensional shapes.
2. Build and draw geometric objects.

Visuals & Resources:
Overhead, pattern blocks, ruler, protractor, glue, construction paper
BLM 4 Colored Triangles

Key Vocabulary:	General Frames:
properties, square, triangle, rhombus, trapezoid, hexagon, characteristic, angles, similar, different	They are similar/different because . . .
HOTS (Higher order thinking questions or tasks):	**Specific Frames:** I created a _____ by . . .
What is something you learned about triangles today?	**Note**: Frames must be explicitly taught and practiced.

Connections to Prior Knowledge/Provide Background Information:
TW ask students to think about what they know about shapes. SW share with a partner at their table and then share out with the whole group. TW remind them of the 4-Corners Vocabulary Chart developed the day before. TW also introduce the new content vocabulary that students will be using in the lesson.

Content Objectives:	Meaningful Activities: Lesson Sequence:	Review/Assessment:
1. SW demonstrate comprehension of two- and three-dimensional geometric shapes by comparing and contrasting triangles based on their sides and angles. **2.** SW demonstrate knowledge of two- and three-dimensional geometric shapes by recalling different geometrical shapes and telling how they are different and the same.	• TW display various triangular shapes and ask questions: What are these shapes? How do you know? How would you describe them? • SW answer questions in small groups then share out with the whole group. Responses should include: They have three sides, three corners, and angles. The sides are straight, etc. • TW distribute triangular figures in different sizes and ask students in their small groups to talk about how they are the same and how they are different, using the general frames: They are similar/different because . . . • TW remind students to use their "tools" to measure sides and angles to identify similarities and differences. • SW share in groups and come to a consensus on a response using the frame: They are similar/different because . . . • TW remind students of the other geometric shapes that they have studied, reviewing key vocabulary from earlier lessons. • TW distribute pattern blocks and talk about what the shapes have in common and how they are different. • SW share in groups and come to consensus on a response using the general frame: They are similar/different because . . .	Student responses Student responses and appropriate use of the general frame Student responses and appropriate use of the general frame

(continued)

SIOP® LESSON PLAN, *Grade 3, Day 2* (continued)

Language Objectives:	Meaningful Activities: Lesson Sequence:	Review/Assessment:
1. SW demonstrate synthesis of two- and three-dimensional geometric shapes by creating geometric shapes using triangles and explaining their creation in a "Find Someone Who" activity using the frame: I created a _____ by . . .	● TW give students BLM 2.1 and ask them to cut out the triangle shapes and create any two of the geometric shapes they have studied. ● SW cut out their triangles and create two geometric shapes and glue them on construction paper. ● SW then be asked to "Find Someone Who" is wearing a similar style and color of shoe to partner up with them. ● SW share their shapes using the specific frame: I created a _____ by . . .	Student-created shapes
	● SW then be asked to "Find Someone Who" is wearing similar style of shirt to partner up with them. ● SW again share their shapes using the specific frame: I created a _____ by . . .	Student shapes and appropriate use of specific frame ***Note:*** *Have students "Find Someone Who" several times. Students will be practicing key vocabulary as they complete their frames.*

Wrap-up: TW have students answer the HOTS question in a "Respond To . . ." SW hand it in on their way out the door.

THINK-ALOUDS for SIOP® Lesson Plan, Grade 3, Day 2

- How will I review for students the measuring tools that they are going to use to identify similarities and differences?

- How do I build on the vocabulary previously taught? New content words must be taught so that students will be able to articulate their understanding of the content concept. What process/functional vocabulary must also be taught in order for students to understand?

- How will I modify the frames for the different language levels in the classroom? All students must be given the opportunity to demonstrate their understanding.

PLANNING POINT for SIOP® Lesson Plan, Grade 3, Day 2

- Note that there are two content objectives for this particular lesson in order to ensure that students are able to review and identify all the geometric shapes they have been introduced to.

SIOP® LESSON PLAN, *Grade 3, Days 3–4*

Key: SW = Students will . . .; TW = Teacher will; HOTS = Higher Order Thinking Skills (questions and tasks)

Unit: Analysis of Geometric Shapes

Lessons 3 and 4: Describing and Drawing Characteristics of Two- and Three-Dimensional Shapes
Grade: 3

NCTM Standard: Analyze characteristics and properties of two- and three-dimensional geometric shapes and develop mathematical arguments about geometric relationships.

Expectations:
1. Identify, compare, and analyze attributes of two- and three-dimensional shapes and develop vocabulary to describe the attributes.
2. Create and describe mental images of objects.

Visuals & Resources:
"Shapes" from *A Light in the Attic*, by Shel Silverstein (Poem)
Geometric shapes (pattern blocks), chart paper, white boards

Key Vocabulary:	*General Frames:*
square, triangle, circle, rectangle, rhombus, trapezoid, hexagon, angles, similar, different	I could see . . . I thought . . . I think it will be the same/different because . . .
HOTS (Higher order thinking questions or tasks): What does visualizing mean? How does visualizing help you learn? When can you use the strategy again? SW be asked to recreate a poem including different shapes.	*Specific Frames:* My illustration is the same because . . . My illustration is different because . . .

Connections to Prior Knowledge/ Provide Background Information:
TW ask students to think about one thing they remember about the geometric shapes they have studied. SW share their responses in Think, Share, Write, and Toss using the frame: "One thing I remember about geometric shapes is . . ."
TW also review the content vocabulary using the examples of the 4-Corners Vocabulary Charts created and posted from Lesson 1.
TW also remind students of what a poem is and of the different poems they have read and studied. TW post examples.

Content Objectives:	*Meaningful Activities: Lesson Sequence:*	*Review/Assessment*
1. SW demonstrate comprehension of two- and three-dimensional geometric shapes by illustrating and describing shapes they recognize in the poem "Shapes."	• TW explain to students that in order to remember new concepts and ideas, they can use a strategy called *visualizing*. • TW explain what the metacognitive strategy is and how it's used, and give examples of when students can use it. • TW explain that in order to help the class continue to develop an understanding of geometrical shapes, she is going to read a poem that includes geometric shapes. She tells them to visualize in their heads how each shape looks as she reads about it. • TW give students a specific example of an idea and a visual representation. • TW read the poem "Shapes" to students once. • TW ask students to share at their tables what geometrical shapes they heard in the poem. • SW share and shout out to the teacher.	

(continued)

Content Objectives:	Meaningful Activities: Lesson Sequence:	Review/Assessment:
	• TW then read the poem once more, this time directing students to visualize what the shapes look like in the poem. • TW then ask students to answer the following questions: What did you see as the poem was read? What did you think as the poem was read? • SW answer in writing using the general frames: I could see . . . I thought . . . • SW turn to a partner and share. • TW choose several students to share with the whole class. • SW then draw on their white boards what the shapes in their heads (square, rectangle, triangle, and circle) looked like. • TW ask another question: Do you think your illustrations are the same as those in the book? Why or why not? • SW will answer the question on their white boards using the general frame: I think it will be the same/different because… • TW put the picture from the book on the overhead and ask students to compare and answer the question: Was your illustration the same or different? Why or why not? • SW use the specific frames to answer: My illustration is the same/different because . . .	Student responses and appropriate use of frames Student illustrations Student responses and appropriate use of frames Student responses and appropriate use of frames

Language Objectives:		
1. SW demonstrate application of the characteristics of two- and three-dimensional geometric shapes by writing a poem including different shapes (rhombus, trapezoid, and hexagon) and orally sharing in small groups.	• TW provide the students with several other shapes (rhombus, trapezoid, and hexagon). • TW remind them of the characteristics of the shapes that they identified in the previous lesson. • In small groups, SW use the shapes to write a new poem on chart paper. • SW also illustrate the poem with the new shapes. • SW share their poems (not the illustration) with the whole class. • SW illustrate on their white boards what they are visualizing in their heads. • SW then share the illustrations that go with the poem they presented. • SW be asked to then compare their illustrations and answer the same question from the preceding frame and share in small groups: Was your illustration the same or different? Why or why not? • SW use the specific frames to answer.	Group poems and illustrations Student responses and appropriate use of frames

Wrap-up: SW be asked to answer the HOTS questions on an index card: What does visualizing mean? How does visualizing help you learn? When can you use the strategy again?

THINK-ALOUDS for SIOP® Lesson Plan, Grade 3, Days 3–4

- Something important that I have thought about and emphasized in this lesson is: How does the language objective ensure that students will apply the concept using the academic language? Writing a poem (the meaningful activity in the language objective) in this lesson is the opportunity for my learners to demonstrate their understanding of the concept as well as their application of the learner strategy (visualizing), which is an essential element of this lesson. I also took into account the different language levels that could be present in my classroom. Can this language objective be modified and adapted for several levels? Yes, students will be working in small groups to reinforce and support their language needs. Another possibility for the language objective could be to complete a cloze poem, in which students only fill in the name of the shapes.

- How much poetry that has been previously taught should I review, since it is assumed that third graders are familiar with poetry? How much understanding of poetry will they need in order to successfully write a poem to meet the Language Objective of the lesson?

- I must also consider that the focus for this lesson is not only to continue to master the mathematical concept but also to develop understanding of a learner strategy. How will I make sure that students clearly understand what I mean by visualizing and that it is a metacognitive strategy that they can continue to use to support their learning?

- How will I group students in a way that will ensure they all participate in the practice and application of the Content and Language Objective? Because students are expected to write in detail, I will have to consider language proficiencies as I group them.

- Students will be asked to write and share several times throughout the lesson using general and specific frames. I must make sure to explicitly teach the frames and model them throughout the lesson to ensure they understand how to effectively use them. What process/functional language will they need to be taught in order to participate in the lesson?

PLANNING POINT for SIOP® Lesson Plan, Grade 3, Days 3–4

- If the majority of the class is made up of students with very limited English language proficiency, the teacher could prepare a cloze poem so that students just have to decide which shape should be included and where it goes in the poem. They will also demonstrate their understanding by illustrating the correct triangle.

SIOP LESSON PLAN, *Grade 3, Day 5*

Key: SW = Students will . . .; TW = Teacher will; HOTS = Higher Order Thinking Skills (questions and tasks)

Unit Focus: Analysis of Geometric Shapes

Lesson 5: Analysis of Geometric Shapes and Application of Content Specific Vocabulary
Grade: 3

NCTM Standard: Analyze characteristics and properties of two- and three-dimensional geometric shapes and develop mathematical arguments about geometric relationships.

Expectations:
1. Identify, compare, and analyze attributes of two- and three-dimensional shapes, and develop vocabulary to describe them.
2. Recognize geometric ideas and relationships and apply them.

Visuals & Resources:

"Grandfather Tang's Story" by Ann Tompert (Book)

Tangram pieces for each student

Key Vocabulary:	General Frames:
square, triangle, rhombus, trapezoid, hexagon, characteristic angles, similar, different **HOTS (Higher order thinking questions or tasks):** Creating examples from the story	Something I learned is . . . It is important because . . . I created a _____ using . . . (what shapes) **Specific Frames:** I used a _____; it has _____ sides. I used a _____; it has _____ angles. I used a _____ and a _____ because they both have _____.

Connections to Prior Knowledge/ Provide Background Information:
SW complete a Graffiti Write in their small groups for the term (*geometric shapes*) (see Planning Points for directions and learning goal).

Content Objectives:	Meaningful Activities: Lesson Sequence:	Review/Assessment:
1. SW demonstrate application of the characteristics of two- and three-dimensional geometric shapes by constructing different examples from a story using geometric shapes.	• TW ask students to post their Graffiti Write posters around the room. • SW stroll around the room and read all of the ideas, words, and illustrations that have been listed for the concept (*geometric shapes*). • TW ask students to think about what they have learned in the Unit and why it is important. • SW answer using the following frames: Something I learned was _____ It is important because . . . . • SW share in a Mix to Music (see Planning Points). • TW then explain that she is going to read one more story for this week's math unit in order for students to demonstrate one last time their understanding of geometric shapes and characteristics. • TW also hand out baggies with tangram pieces to each student.	Ideas, words, and statements on student-generated posters Student answers and appropriate use of frames

Content Objectives:	Meaningful Activities: Lesson Sequence:	Review/Assessment:
	• TW share what they are and model on the overhead how they can be used to create different shapes. • TW remind students that throughout the unit they have used different resources and tools to create shapes and explain their characteristics. SW refer to those examples as references to work through today's lesson. • TW introduce the story and read it in parts so that students can take the time to create the different illustrations. The different parts of the story will be reproduced on overheads so that students can follow along and have a visual of what they are being asked to recreate. • TW model how to create the first illustration from the story and have students follow along. • SW talk at their tables about the different pieces they used and how they are similar and different. • TW continue to read the story. • SW create the different illustrations (characters) from the story. • TW stop at appropriate points and have students share their creations, explaining what shapes they put together using the following frame: I created a _____ using . . . (what shapes). • TW continue until the story is read.	Teacher observation and monitoring as students are creating the different examples from the story. Are they able to manipulate shapes in a way to effectively create the examples? Student creations and appropriate use of frames

Language Objectives:		
1. SW demonstrate evaluation of the characteristics of two- and three-dimensional geometric shapes by comparing different shapes and explaining their characteristics using key vocabulary.	• TW give each group a copy of the story and ask them to read through it as a group. • TW ask them to each recreate their favorite example from the story using their tangram pieces. • SW create and take turns sharing their example and explaining to the group what they have created and what shapes they had to use in order to do so. • SW then draw the example on a blank sheet of paper and complete the following frames to explain the shapes they used and why they chose those shapes using the frames: I used a _____; it has _____ sides. I used a _____; it has _____ angles. I used a _____ and a _____ because they both have .	Student examples and appropriate use of frames in writing and when orally sharing on their sheet that will be turned in to the teacher

Wrap-up: SW will be asked to go back to their Graffiti Write chart paper with their group and take one minute to add anything new that they now know.

PLANNING POINTS for SIOP® Lesson Plan, Grade 3, Day 5

● A Graffiti Write is an activity that allows students to quickly share their ideas and thoughts on a particular topic or concept. In small groups, they are given a piece of chart paper with the topic/concept listed in the middle of the paper. Each student is given a marker and, at the same time, all students list words, ideas, questions, or an illustration of what they know about the topic in two to three minutes. In this lesson, the topic *geometric shapes* was written on chart paper and the Graffiti Write was used as an opportunity to tap into students' prior knowledge and experiences. It was also an opportunity for the teacher to assess what the students remembered and their level of understanding.

● Used at the beginning of this lesson, Mix to Music is a fun activity that involves all students sharing their sentence frames with one another. Using a frame takes pressure off students to come up with an original, correct sentence. Mix to Music is a structured opportunity for students to share formulated responses. All too often, teachers ask students to formulate responses, but only have a few students share. Those who usually share are students who aren't struggling with the concept or language, and they are usually not English learners. When we take the traditional approach of "Who would like to share?" we give struggling learners as well as ELs the opportunity to opt out of the learning process. Structures like Mix to Music ensure that everyone shares and practices using language. The first step is to have students stand with their responses. The teacher plays music and has students mix around the room. When the music stops, students grab a partner closest to them and take turns sharing their responses. When the music starts again, the process is repeated. Repeat several times so that students get ample opportunity to use all four language domains (reading, writing, listening, and speaking).

THINK-ALOUDS for SIOP® Lesson Plan, Grade 3, Day 5

● How will I address any gaps I observe as students create examples from the story using geometric shapes? I think that having students respond continuously through the lesson with specific frames allows them to articulate their understanding—or lack thereof—so that I can evaluate students' understanding and pinpoint what is still missing.

● I always need to consider the effectiveness of partner match-ups. In this lesson I'm not sure that my plan will work well since students are randomly partnered up. Will some students be at a disadvantage in terms of using language or not hearing good language modeled for them because of the different language proficiency levels in the classroom?

Concluding Thoughts

In this unit, notice how language practice is infused into math lessons by providing students with sentence frames. The frames structure student responses and scaffold their oral practice. Many English learners—even those at advanced levels of proficiency—benefit from having their language production scaffolded for them, which reduces the cognitive load linguistically, allowing for focused attention on the math concept.

For some math teachers, using reading books in math lessons is not their usual practice. Books were used in this unit to facilitate understanding of math concepts (geometric shapes) and to develop language by providing more exposure to the vocabulary terms.

The level of detail in the SIOP® lesson plans presented in this unit was quite extensive. It was necessary for us to describe the exact steps in each lesson and to make the activities and processes comprehensible for readers. You undoubtedly will have your own shorthand and symbols to represent much of what was described in narrative form in our lessons. Regardless of the level of detail you use, it is helpful to keep the SIOP® protocol handy as you write lesson plans so that you make sure that all components are present in your lessons. And, as mentioned previously, the more practice you have in developing SIOP® lessons and units, the easier and more natural the process becomes.

Lesson and Unit Design for SIOP® Mathematics Lessons

Grades 6–8

Araceli Avila

Math Unit, Grades 6–8

Overview of the Unit

As you know, the middle school years mark a transitional phase for students. During this time period, mathematical content and language become more complex and abstract than what was taught in the elementary grades. Furthermore, one can't help but notice that middle school students begin to encounter issues and changes brought on by adolescence. If we factor in lack of English proficiency, middle school mathematics can be a difficult and stressful subject for English learners. Consequently, in order to accelerate the acquisition of English and to engage the interest of ELs in the content and language of middle school mathematics, we need to pay attention to the type of environment we create in our classrooms. It should be an exciting place to be, one that is language rich, safe, supportive, and yet challenging.

In our experience, English learners are most successful when lessons not only consider the national and state content standards but also the language and social-emotional needs of ELs. We have found that students need the opportunity to practice the new content knowledge concretely and use the language skills of reading, writing, listening, and speaking to acquire the academic language of mathematics. This in turn will promote the understanding and usefulness of mathematics. In the following seventh grade unit, you will find how the eight SIOP® components and features were incorporated into the lessons not only to allow students to practice and learn the rigorous math content concepts and academic language of adding and subtracting integers but also to captivate their interest and motivate them to become learners of mathematics.

This unit of study is guided by the questions: What is an integer? and How can the rules of adding and subtracting integers be applied to solve problems from real-world situations? By the end of the unit, students will have had sufficient instruction and multiple exposures to the academic language needed to answer these questions. The blackline masters (BLMs) that support the unit are in Appendix C.

Throughout the unit, you will also notice that there are Planning Points and Think-Alouds. As experienced SIOP® teachers, we have included in the lessons our own thought processes for planning SIOP® lessons that will result in the high-quality teaching described above. The purpose of the Planning Points is to highlight the process of planning a SIOP® unit and to clarify any questions you may have about each specific lesson plan. The Think-Alouds are questions we asked ourselves or issues that arose throughout the SIOP® unit planning process. We thought it would be helpful to include them as a possible resource for your own planning.

We hope that you will use the models we have provided in this chapter as a guide for your own planning and teaching. You may find that some of the lessons will require more or less time than what is allotted here, depending on your students. That is fine; the lessons are designed to show how SIOP® features are integrated into math lessons over the course of a unit of study. We encourage you to adapt them for your own students' needs.

SIOP® Planning Flow Chart

Grade Level: 7   **Subject:** Math
Unit Concept: (Big Idea) Adding and   **Approximate Time Involved:** 5 days
                    Subtracting Integers

Lesson Focus Day 1:
The Meaning of Integers

Content Objectives:
SW represent integers in real-life situations.

Language Objectives:
SW discuss and list real-life situations represented by integers.
SW predict how the temperature of the water will change by completing the sentence starter:
    If you add ice to the glass, I predict the temperature will . . .
    If you add hot water to the water, I predict the temperature will . . .

Reading/Writing/Discussion Activities:
4-Corners Vocabulary; Sentence Starters; Cloze Sentences; Think, Pair, Share; Brainstorming; Who Is Colder? Card Game

Lesson Focus Day 2:
Adding Integers

(continued)

Content Objectives:
SW model addition of integers using two-color counters and number lines.
SW discover rules for adding and subtracting integers.

Language Objectives:
SW compare and contrast positive and negative integers by completing a Venn Diagram.
SW write at least one pattern you discovered about adding integers and share findings with a partner by participating in a Conga Line Activity.
    A pattern I discovered is . . . .

Reading/Writing/Discussion Activities:
4-Corners Vocabulary; Sentence Starters; Think, Pair, Share; Conga Line; Simultaneous Round Table

Lesson Focus Day 3:
Subtracting Integers

Content Objectives:
SW model subtraction of integers using two-color counters.
SW discover rules for adding and subtracting integers.

Language Objective:
SW review vocabulary words by participating in "Vocabulary Alive" activity.
    The word is _____ and it looks like . . .
SW in small groups, discuss patterns that they discovered about subtracting integers and write three patterns on an index card.
    Three patterns we discovered are . . .

Reading/Writing/Discussion Activities:
Vocabulary Alive; 4-Corners; Sentence Starters; Fun with Integers Game

Lesson Focus Day 4: Applying the Rules for Adding and Subtracting Integers

Content Objectives: SW apply the rules for adding and subtracting integers to real-life situations.

Language Objectives: SW write a real-life situation involving integers and have other students solve by participating in "You Are the Teacher" activity.

Reading/Writing/Discussion Activities: Vocabulary Alive; You Are the Teacher; Think, Pair, Share

SIOP® LESSON PLAN, *Grade 7, Days 1–2 The Meaning of Integers*

Class/Subject Area(s): Math
Unit/Theme: Adding and Subtracting Integers

Grade Level: 7
Lesson Duration: 90 minutes

NCTM Standards

Content	Process
Number and Operations • Develop meaning for integers and represent and compare quantities with them. • Understand the meaning and effects of arithmetic operations with integers. • Develop and analyze algorithms for computing with integers and develop fluency in their use.	☐ Problem Solving ☐ Reasoning & Proof ☐ Communication ☐ Connections ☐ Representations

Content Objective(s):
SW represent integers in real-life situations.

Language Objective(s):
- Day 1: SW discuss and list real-life situations represented by integers.
- Day 2: SW predict how the temperature of the water will change by completing the sentence starter:

    "If you add ice to the glass, I predict the temperature will . . ."
    "If you add hot water to the glass, I predict the temperature will . . ."

Key Vocabulary:		*Supplementary Materials:*	

Content Vocabulary
- Integers
- Positive integers
- Negative integers
- Number line
- Opposites
- Absolute value
- +/–

Functional Vocabulary
- Compare
- Order

- Masking tape for number line
- Integer dollar cards, BLM 9
- Decks of cards
- Lab sheet, 10
- 4-Corners Vocabulary Charts
- Instructions for Who Is Colder? Card Game, BLM 11

- Thermometers
- Cups
- Water (room temperature and hot water)
- Ice cubes
- Chart paper for brainstorming activity
- Index cards

SIOP Features:

Preparation
- ✗ Adaptation of content
- ✗ Links to background
- ✗ Links to past learning
- ✗ Strategies incorporated

Scaffolding
- ✗ Modeling
- ✗ Guided practice
- ✗ Independent practice
- ✗ Comprehensible input

Grouping Options
- ✗ Whole class
- ✗ Small groups
- ✗ Partners
- __ Independent

Integration of Processes
- ✗ Reading
- ✗ Writing
- ✗ Speaking
- ✗ Listening

Application
- ✗ Hands-on
- ✗ Meaningful
- ✗ Linked to objectives
- ✗ Promotes engagement

Assessment
- ✗ Individual
- ✗ Group
- __ Written
- __ Oral

Lesson Sequence:

1. Ask students to work in small groups to list the types of numbers they have worked with in previous units or school years. (Examples: whole and natural numbers, fractions, decimals, irrational numbers like pi.) Have a whole group discussion about the types of numbers they have learned about in the past. Tell them they will learn about a new type of number called *integer*.

2. Activate students' prior knowledge by asking them to think of a time when they borrowed money. On an index card, tell them to complete the following sentence frame: "One time I borrowed money was . . ." Have students participate in a Think, Pair, Share.

3. Engage students in a discussion about borrowing money. Ask questions like: Who does the money belong to? Why do you have to pay borrowed money back? Is there a symbol or number in mathematics that can represent borrowed money? If yes, what is that symbol or number?

4. Explain to students that negative integers are used in mathematics to represent borrowed money. Use a number line to represent on which side of the number line borrowed money exists. Also explain to them where earned money exists on the number line. Pass out the borrowed/earned integer cards, BLM 9. Tell each student to pretend the card they have represents money they borrowed or earned. Instruct them to turn to a partner and state whether their card represents borrowed or earned money and to determine where they belong on the number line. They can use the following cloze sentence: "I borrowed/earned __ dollars; I belong on the __ side of zero." Have students compare and order themselves on the number line based on the card they received.

(continued)

SIOP® LESSON PLAN, *Grade 7, Days 1–2 The Meaning of Integers* (continued)

5. Construct a number line on the overhead and plot each of the students' card numbers. Point out the borrowed $10 and the earned $10. Ask them to turn to a partner and determine how these numbers are the same and how they are different. Allow students to share their responses with the whole group. Introduce the concept of opposite numbers and absolute value.

6. Explain to students that this unit will focus entirely on integers, their meaning, and how to add and subtract them. Read the content and language objectives to students.

7. Introduce content and functional vocabulary verbally and pictorially. Allow students to process the information by filling out a 4-Corners Vocabulary Chart for each term.

8. Tell students that many real-life situations can be represented with integers. In small groups have students brainstorm situations other than borrowing money that can be represented by integers. Ask one student from each group to give you an example. Write their responses on chart paper.

9. Explain to students that temperature is a real-life situation that is represented by positive and negative integers. If available, show a video related to temperature and integers.

10. Show students two glasses of water. Tell students the temperature of the water is at room temperature. Ask them to predict the temperature of the water. Use a thermometer to determine the temperature of the water in both glasses. Add hot water to one of the glasses of water and ice cubes to the second glass. Ask students to predict how the temperature of each glass will change. State to students that they will work in small groups to determine by how many degrees the temperature of the water changed.

11. Place students in groups of two or three. Provide each group two glasses of water at room temperature, one thermometer, and a lab sheet, BLM 10. Have each group complete the lab sheet.

12. Once all groups are done, have each group share their findings with the whole class. At this point make a connection between the words *decrease* and *increase* in temperature to the symbolic representation of integers. For example, a decrease of 4 degrees can be represented by -4 degrees. Have students go back to the lab sheet and include the symbolic representation for the change in temperature.

13. Have students review positive and negative integers by playing a game of "Who Is Colder?" BLM 11.

14. Wrap up today's lesson by having students participate in a Number 1-3 for Self-Assessment of lesson's objectives.

Reflections:
After teaching the lesson, the teacher reflects on what worked, what did not work, and what revisions, additions, and/or deletions need to be made.

PLANNING POINT for SIOP® Lesson Plan, Grade 7, Days 1–2

● I would suggest that this lesson be taught over a two-day span. Notice that the content objective does not change. However, language objectives must be carefully selected and planned to ensure that students not only practice reading, writing, speaking, and listening skills but also master the content objective. For example, the language objective for Day 1 serves a dual purpose. First, it helps students brainstorm about their experiences with integers while allowing them to use a variety of language skills. Second, it assists the teacher in assessing students' prior knowledge. If the teacher notices that students have little or no prior knowledge about integers, then the teacher knows to build background. Consequently, building background aids ELs in mastering the content objective.

THINK-ALOUDS for SIOP® Lesson Plan, Grade 7, Days 1–2

- What questions can I ask to assess students' prior knowledge about positive and negative integers? What is an experience all of my students might have in common?

- What are the essential key terms students must know to be successful in this unit/lesson? How will I ensure key terms are not only introduced by me but also processed, used and reviewed by students? Many times students forget to use the terminology when discussing with their partners. How can I make sure terms are being used consistently by students?

- Many times EL students report that teachers are not clear when giving instructions. Therefore, how will I make sure my explanation of academic tasks is clear? Will I write the step-by-step instructions on the board? Will I state the instructions and then have a couple of students restate the instructions to the class?

- Refer to Chapter 3 for a full discussion of 4-Corners Vocabulary and Chapter 4 for Number 1–3 for Self-Assessment of Objectives.

SIOP® LESSON PLAN, *Grade 7, Day 3 Adding Integers*

Class/Subject Area(s): Math   **Grade Level:** 6–8
Unit/Theme: Adding and Subtracting Integers   **Lesson Duration:** 60 minutes

NCTM Standards

Content	Process
Number and Operations • Develop meaning for integers and represent and compare quantities with them. • Understand the meaning and effects of arithmetic operations with integers. • Develop and analyze algorithms for computing with integers and develop fluency in their use.	☐ Problem Solving ☐ Reasoning & Proof ☐ Communication ☐ Connections ☐ Representations

Content Objective(s):
SW model addition of integers using two-color counters and a number line.
SW discover rules for adding and subtracting integers.

Language Objective(s):
SW compare and contrast positive and negative integers by completing a Venn Diagram.
SW write about at least one pattern he/she discovered and share findings with a partner by participating in a Conga Line Activity.
"A pattern I discovered is . . . ."

(continued)

SIOP® LESSON PLAN, *Grade 7, Day 3*
Adding Integers (continued)

Key Vocabulary:		Supplementary Materials:

Content Vocabulary
- Zero pair
- Equation or number sentence

Functional Vocabulary
- Compare
- Contrast
- Combine

- Integer Venn Diagram, BLM 3.1
- Weather News Transparency BLM 3.2
- Adding Integers Lab Sheet BLM 3.3
- Simultaneous Round Table Activity Sheet BLM 3.4
- Response boards and dry erase markers
- Two-color counters
- Index cards
- Outcome sentences

SIOP® Features:

Preparation
- ✗ Adaptation of content
- ✗ Links to background
- ✗ Links to past learning
- ✗ Strategies incorporated

Scaffolding
- ✗ Modeling
- ✗ Guided practice
- ✗ Independent practice
- ✗ Comprehensible input

Grouping Options
- ✗ Whole class
- ✗ Small groups
- ✗ Partners
- ✗ Independent

Integration of Processes
- ✗ Reading
- ✗ Writing
- ✗ Speaking
- ✗ Listening

Application
- ✗ Hands-on
- ✗ Meaningful
- ✗ Linked to objectives
- ✗ Promotes engagement

Assessment
- ✗ Individual
- ✗ Group
- ✗ Written
- ___ Oral

Lesson Sequence:

1. Show a picture of a number line and a thermometer, and ask students to think quietly about what they learned in the previous lesson. Have them participate in a Think, Pair, Share. Discuss with the whole class.

2. Group students in pairs and have students compare and contrast positive and negative integers by completing a Venn diagram, BLM 12. Once all groups are done, have pairs form small groups of four and compare their Venn diagrams with one another. Discuss with the whole class.

3. Provide a response board and dry erase marker to each student. Display the Weather News Transparency, BLM 12. Ask students to read the problem and use their response boards to find the solution. Check their answers by having them raise their boards. Ask several students to explain how they got their answer. Did you draw any pictures? Why? What do the pictures represent? What operation did you use? How did you know you had to add? Is the final temperature positive or negative?

4. Introduce the lesson's content and language objectives.

5. Pose the question, "What would be the final temperature if at 7:00 AM the temperature was 0 degrees C, then later in the morning the temperature would rise 3 degrees C and in the evening the temperature dropped 3 degrees C?" Have students share their answers with a partner and then have a whole class discussion. Model this problem using two-color counters. (Instruct students that the yellow side of the counter represents a +1 and the red side represents a −1.) Tell students that the yellow and red counters are opposite integers. What happens when you combine a yellow counter and a red counter? (Opposite integers cancel each other.) Tell students that opposite integers that cancel each other are called *zero pairs*. Model this problem using a number line. Ask students, "How are these two models, two-color counters and number line, the same and different?"

6. Introduce the terms *zero pair* and *equation* verbally and pictorially. Have students process the meaning of the words by constructing a 4-Corners Vocabulary Chart. (Refer to Ch. 3 for a description.)

7. Provide each student with the Adding Integers Lab Sheet 1, BLM 14. Model the first three examples for them. Have students work in pairs to complete the rest of the lab sheet.

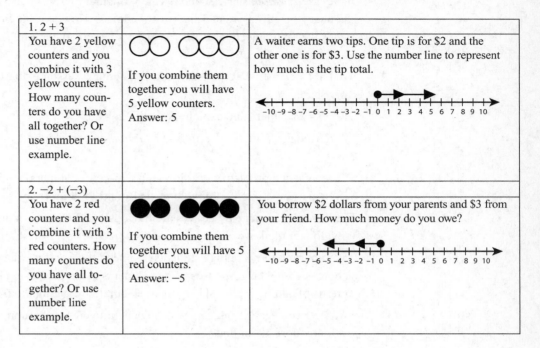

1. 2 + 3		
You have 2 yellow counters and you combine it with 3 yellow counters. How many counters do you have all together? Or use number line example.	If you combine them together you will have 5 yellow counters. Answer: 5	A waiter earns two tips. One tip is for $2 and the other one is for $3. Use the number line to represent how much is the tip total.
2. −2 + (−3)		
You have 2 red counters and you combine it with 3 red counters. How many counters do you have all together? Or use number line example.	If you combine them together you will have 5 red counters. Answer: −5	You borrow $2 dollars from your parents and $3 from your friend. How much money do you owe?

8. In pairs, have students discuss patterns they discovered about adding integers. Ask each student to write the patterns on an index card. (Sentence Starter: "One pattern I discovered . . .") Students must use the unit's vocabulary to describe their patterns. Have students share their patterns by participating in a Conga Line activity. (Examples of patterns: You add two positive integers and the answer is positive; you add two negative integers and the answer is negative; you subtract a positive and a negative integer and keep the sign of the largest numeral.) Afterwards, have a whole class discussion.

9. To review today's content concepts, have students participate in a Simultaneous Round Table activity sheet, BLM 15. Place students in groups of 4. Provide each student with a copy of the Adding Integers Simultaneous Round Table activity sheet. In each rotation, a student will complete one of the quadrants. When the sheet is complete, the activity sheet goes back to its owner. The student is then responsible to determine whether the answers are correct (NOTE: For continued practice with adding integers, assign homework from the district adopted textbook, practicing the same skill.)

10. To wrap up the lesson, ask students to participate in outcome sentences: "I learned . . . , I liked . . . , I wonder . . ." Review content and language objectives.

Reflections:
After teaching the lesson, the teacher reflects on what worked, what did not work, and what revisions, additions, and/or deletions need to be made.

PLANNING POINTS for SIOP® Lesson Plan, Grade 7, Day 3

- In this lesson I want to provide students with an opportunity to use and develop their writing skills. Remember to differentiate the writing portion of this activity based on students' language proficiency levels. Students may draw a picture or use their native language to explain their thinking to a classmate who speaks the same native language.

- This lesson provides students ample opportunities to practice their language skills of reading, writing, listening, and speaking. For instance, students complete a Venn diagram, participate in Think, Pair, Share, Conga Line, and Simultaneous Round Table, use sentence starters, and define vocabulary terms. When selecting the lesson's language objectives, we had an array of choices. However, the SIOP® Model recommends we limit objectives to one or two per lesson. Other language objectives we could have chosen are:

- SW tell a partner what they learned in yesterday's lesson by participating in a Think, Pair, Share.

- SW read a problem and explain how they solved it.

- SW define the words *zero pair* and *equation* by completing a 4-Corners Vocabulary Chart.

- SW represent a problem's solution with two-color counters, number line, and equation by participating in a Simultaneous Round Table Activity.

- The lesson's language objectives are obviously different from the ones written above. Why do you think we chose the lesson's objectives over these?

- It is important to encourage students to extend their oral language repertoire by elaborating on their responses. When students begin discussing the emerging patterns of adding integers, help them elaborate by asking them questions like:

  - When two negative integers were added, did you add or subtract? Why? Can you tell me more about that?

  - Was the answer positive or negative? How did you know?

  - How does adding two red counters differ from adding two yellow counters? What else?

  - Did you add or subtract when you combined different signed integers/colored counters? Why or why not? Was the answer positive or negative? Why? Does anyone think differently?

- It's important that the homework I assign is extra practice of a skill I have already taught and that students have practiced with my supervision. The goal of homework is reinforcement of a skill students are comfortable with—not having them delve into new material or skills they barely know.

- Keep in mind that students might not easily recognize the patterns of adding and subtracting integers. ELs might struggle with both identifying patterns and formulating responses in English. Therefore, use prompting techniques and provide sufficient wait time for students to respond.

THINK-ALOUD for SIOP® Lesson Plan, Grade 7, Day 3

- Does this lesson allot sufficient time to meet content and language objectives? If the pacing of the lesson is too fast for students based on their educational background and language proficiency skills, I should consider spending two days on this lesson.

- Refer to Chapter 4 for Conga Line Activity.

Class/Subject Area(s): Math   **Grade Level:** 7
Unit/Theme: Adding and Subtracting Integers   **Lesson Duration:** 60 minutes

NCTM Standards

Content	Process

Number and Operations
- Develop meaning for integers and represent and compare quantities with them.
- Understand the meaning and effects of arithmetic operations with integers.
- Develop and analyze algorithms for computing with integers and develop fluency in their use.

- ☐ Problem Solving
- ☒ Reasoning & Proof
- ☒ Communication
- ☒ Connections
- ☒ Representations

Content Objective(s):
SW model subtraction of integers using two-color counters.
SW discover rules for adding and subtracting integers.

Language Objective(s):
SW review vocabulary words by participating in Vocabulary Alive.
   "The word is _____ and it looks like this . . ."
In small groups, SW discuss patterns they discovered about subtracting integers and write three patterns
   on an index card.
   "Three patterns we discovered are"

Key Vocabulary:		Supplementary Materials:

Content Vocabulary
- Additive inverse
- Vocabulary covered in previous lessons

Functional Vocabulary
- Remove

- Two-color counters
- Subtracting Integers Lab Sheet BLM 16
- Fun with Integers Instructions BLM 17
- Fun with Integers Recording Sheet BLM 18
- Spinner
- Deck of cards
- Chart paper

SIOP® Features:

Preparation
- **X** Adaptation of content
- **X** Links to background
- **X** Links to past learning
- **X** Strategies incorporated

Scaffolding
- **X** Modeling
- **X** Guided practice
- **X** Independent practice
- **X** Comprehensible input

Grouping Options
- __**X** Whole class
- **X** Small groups
- **X** Partners
- **X** Independent

Integration of Processes
- **X** Reading
- **X** Writing
- **X** Speaking
- **X** Listening

Application
- **X** Hands-on
- **X** Meaningful
- **X** Linked to objectives
- **X** Promotes engagement

Assessment
- **X** Individual
- **X** Group
- **X** Written
- **X** Oral

Lesson Sequence:

1. Review unit's vocabulary words by participating in Vocabulary Alive.

2. Ask students, "Who remembers what we learned yesterday?" Review patterns students discovered in the previous lesson. (Use the unit's vocabulary to restate students' patterns.) Construct a T-Chart

(continued)

SIOP LESSON PLAN, *Grade 7, Day 4*
Subtracting Integers (continued)

on chart paper. Label one side, Adding Integers, and the other side, Subtracting Integers. Write down the students' addition patterns. Tell students today's lesson will focus on subtracting integers with two-color counters.

3. Introduce content and language objectives.

4. Provide each student with two-color counters and the Subtracting Integers Lab Sheet, BLM 16. Model the first three problems for students. The first three examples will lead in to the introduction of the concept of *Additive Inverse*. Have students construct a 4-Corners Chart for the new word.

1. $-3 - 4$	
You have 3 red counters and you need to remove 4 yellow counters. There are no yellow counters to remove. Add 4 zero pairs to -3 and now remove 4 yellow counters. How many counters are left?	● ● ● ○ ○ ○ ○ ● ● ● ● If you remove 4 yellow counters you are left with 7 red counters. Answer: -7

5. Allow students to work in pairs to finish the lab sheet. Have them pay attention to emerging patterns.

6. Group students in triads, and ask them to write three patterns that surfaced from subtracting integers with two-color counters. Collect index cards and read each group's patterns to the whole class. Write subtraction patterns on the T-Chart. Ask students to discuss in small groups if there are any similarities or differences between the patterns for adding and subtracting integers. Have one person from each group share their findings.

7. Formalize the rules for adding and subtracting integers by having a whole group discussion.

Addition Rules	Subtraction Rules
• Add two positive integers and sum is positive $(+ + +;$ add, keep $+$ sign) • Add two negative integers and sum is negative $(- + -;$ add, keep $-$ sign) • Subtract a positive integer and negative integer and keep the sign of the largest numeral. $(- + +$ or $+ + -;$ subtract, keep sign of largest numeral)	Change subtraction to addition by applying the additive inverse. Then follow the addition rules.

8. In order for students to practice applying the rules of adding and subtracting integers, have them play a game of Fun with Integers (BLM 17). Give each group of students a recording sheet, BLM 18, and have them follow the game's instructions. (NOTE: For continued practice with subtracting integers, assign homework from the district adopted textbook.)

9. Review content and language objectives.

Reflections:
After teaching the lesson, the teacher reflects on what worked, what did not work, and what revisions, additions, and/or deletions need to be made.

PLANNING POINTS for SIOP® Lesson Plan, Grade 7, Day 4

- For the Vocabulary Alive activity, it is important to display vocabulary terms on a word wall, chart paper, or PPT. At the end of Vocabulary Alive consider quizzing students by acting out each gesture and having students write the corresponding words. When taking the quiz, ELs benefit from having the list of words available.

- When compared to addition of integers, subtraction is a more difficult concept to grasp. Therefore, when modeling subtraction with two-color counters, consider using the word *remove* instead of *subtract*. Have students explain their reasoning as they subtract integers. Have them explain why addition needs to be applied to solve $-8 - 3$.

THINK-ALOUDS for SIOP® Lesson Plan, Grade 7, Day 4

- Students are usually told to memorize the rules for adding and subtracting integers. What meaningful strategy can I use so that students can remember the rules? Can mnemonics, Total Physical Response, etc. be applied?

- How will I make sure students are using their higher order thinking skills? Does the lesson incorporate both questions and tasks that promote the use of higher order thinking? How will I make sure my lower level language students have an opportunity to answer these questions?

- Refer to Chapter 4 for Vocabulary Alive.

SIOP® LESSON PLAN, *Grade 7, Day 5 Applying Rules for Adding and Subtracting Integers to Real-Life Situations*

Class/Subject Area(s): Math
Unit/Theme: Adding and Subtracting Integers

Grade Level: 7
Lesson Duration: 60 minutes

NCTM Standards

Content	Process
Number and Operations • Develop meaning for integers and represent and compare quantities with them. • Understand the meaning and effects of arithmetic operations with integers. • Develop and analyze algorithms for computing with integers and develop fluency in their use.	☒ Problem Solving ☒ Reasoning & Proof ☒ Communication ☒ Connections ☒ Representations

Content Objective(s):
SW apply the rules for adding and subtracting integers to real-life situations.

Language Objective(s):
SW write a real-life situation involving integers and have other students solve by participating in a "You Are the Teacher" activity.

Key Vocabulary:		*Supplementary Materials:*
Content Vocabulary • Unit's vocabulary	**Functional Vocabulary** • Ascend • Descend	• Where Is the Submarine? BLM 19 • Applying Integers Lab Sheet, BLM 20 • Two-color counters (optional) • Chart paper • Marker

SIOP® Features:

Preparation
✗ Adaptation of content
✗ Links to background
✗ Links to past learning
✗ Strategies incorporated

Scaffolding
✗ Modeling
✗ Guided practice
✗ Independent practice
✗ Comprehensible input

Grouping Options
✗ Whole class
✗ Small groups
✗ Partners
✗ Independent

Integration of Processes
✗ Reading
✗ Writing
✗ Speaking
✗ Listening

Application
__ Hands-on
✗ Meaningful
✗ Linked to objectives
✗ Promotes engagement

Assessment
✗ Individual
__ Group
✗ Written
__ Oral

(continued)

SIOP® LESSON PLAN, *Grade 7, Day 5 Applying Rules for Adding and Subtracting Integers to Real-Life Situations* (continued)

Lesson Sequence:

1. Review the unit's vocabulary terms by going over the gestures from Vocabulary Alive and reviewing the rules for adding and subtracting integers.

2. Introduce content and language objectives.

3. Access students' prior knowledge by asking, "What is a submarine?" Have them share their thoughts with a partner and then have a whole group discussion. If there is a need to build background, show pictures of submarines or play a video. (NOTE: A short video can be found at http://www.youtube.com/watch?v=3MUBgG6rzO8.)

4. Have students work in pairs on the "Where Is the Submarine?" problem, BLM 5.1. Students are responsible for coming up with a solution and explaining how they solved the problem. Once all groups are done, have each group share their solution with the whole group. On chart paper, scribe all the different methods for solving the problem.

5. Group students in pairs or triads and have them work on the Applying Integers lab sheet, BLM 5.2.

6. Now that students have applied the rules of adding and subtracting integers, have them write a word problem for other students to solve by participating in "You Are the Teacher." Group students in triads and post chart paper around the room for each group. Assign each group to write a word problem involving integers. Once all groups are done writing, one student from each group stays in their assigned group and the rest rotate to the next group. The student who stayed behind reads the problem to his/her new group members. As a group, they need to use two methods to solve the problem. Identify a new teacher and have the rest rotate. In this new rotation, the student who stayed reads the problem and explains how his/her previous team solved the problem. The group's responsibility is to determine whether the two solutions are correct.

7. Assess students' comprehension of adding and subtracting integers by having them individually solve the following problems:

Problem 1	In Minneapolis, MN the temperature was −14°F in the morning. If the temperature dropped 7°F, what is the current temperature?	Method 1	Method 2
Problem 2	In the Sahara Desert, one day it was 127°F. In the Gobi Desert a temperature of −30°F was recorded. What is the difference between these two temperatures?	Method 1	Method 2

8. As a wrap-up activity, have students reflect on the following question: "Why are integers useful?" Close the lesson by reviewing content and language objectives.

Reflections:

After teaching the lesson, the teacher reflects on what worked, what did not work, and what revisions, additions, and/or deletions need to be made.

PLANNING POINTS for SIOP® Lesson Plan, Grade 7, Day 5

- NCTM recommends that all students communicate their mathematical thinking through coherent and clear explanations to their classmates and teachers. Therefore, it is important to provide ample opportunities for ELs to practice the language skills of reading, writing, speaking, and listening. Students might struggle with formulating their responses in English, but we need to remember that practice makes perfect.

- Conceptual understanding of mathematical concepts is a key goal of national and content standards. The SIOP® component of Practice & Application also supports this goal. SIOP® Feature 20 states that students need to practice new content knowledge through the use of concrete materials and manipulatives. This feature allows ELs to develop conceptual understanding of key concepts and practice with the language of mathematics. Lessons 1–3 were designed to meet the goal of conceptual understanding. In Lesson 4, students are applying what they learned about integers to real-life situations. This lesson addresses SIOP® Feature 21 because students need to apply their understanding of integers and use appropriate math and functional language to solve meaningful problem situations. Keep in mind that continued practice and application will lead students to develop conceptual understanding and fluency in the use of algorithmic operations.

 THINK-ALOUDS for SIOP® Lesson Plan, Grade 7, Day 5

- How will I consider the language needs of ELs who might be in the initial stages of second language acquisition? How will I differentiate this lesson for them?
- If students are still struggling with the application and use of the algorithm, should I allow students to use two-color counters, number lines, or calculators to solve problems?
- Refer to Chapter 4 for Vocabulary Alive and You Are the Teacher.

Concluding Thoughts

Each grade level comes with its own issues and challenges. It is important for middle school students to find the classroom to be a safe, supportive environment and one that challenges them to meet high academic standards. We believe you will find that this unit provides a balance between rigorous standards-based content with challenging content and language objectives and teaching that is designed to promote success for all students. In fact, some of the lessons may seem too difficult for your English learners, but differentiating instruction through grouping, by modifying the activities and assignments, by providing frames, and by allowing time for more practice will make the lessons appropriate for students at all levels of language proficiency. Please refer back to Chapters 3 and 4 to see how activities can be modified to accommodate students at the beginning, intermediate, and advanced high levels of language proficiency.

Lesson and Unit Design for SIOP® Mathematics Lessons

Grades 9–12

Araceli Avila

Math Unit, Grades 9–12

Overview of the Unit

In order to meet their school's rigorous graduation requirements, English learners must learn the standards-based content and academic language of high school mathematics. Further, nearly half of the states in the United States have high school exit exams, in addition to No Child Left Behind (NCLB) testing requirements. For those states that have high school exit exams, only three have specific alternatives geared for ELs. These requirements have resulted in an increasing number of ELs failing to attain a graduation diploma.

A key barrier in English learners' quest for a graduation diploma is their lack of academic English proficiency. As teachers of mathematics, we can assist ELs in meeting

graduation requirements by helping them develop both mathematical understanding and English proficiency. The following ninth grade unit was designed with this in mind. The lessons provide ELs ample opportunities to practice and apply the content concepts of transforming linear, quadratic, and exponential parent functions while engaging them in mathematical conversations that will facilitate the development of the academic English they need to pass high-stakes assessments. By the end of this unit, students will be able to answer, "What effects do transformations cause on linear, quadratic, and exponential parent functions?"

The lesson on Comparing and Contrasting Functions in the unit may seem to be more appropriate for middle school students rather than high school students. However, I intentionally took a complex concept—representing a problem situation in multiple ways—and presented it in a fairly straightforward, simple way to accommodate students who may have low levels of English proficiency or gaps in their education. This lesson is geared to the level of educational background and language proficiency of my students; you may want to increase its rigor or elevate the complexity of the problem situation to meet the needs of your own students.

Throughout the unit, you will also notice that there are Planning Points and Think-Alouds. As experienced SIOP® teachers, we have included in the lessons our own thought processes for planning SIOP® lessons that will result in the high-quality teaching described above. The purpose of the Planning Points is to highlight the process of planning a SIOP® unit and to clarify any questions you may have about each specific lesson plan. The Think-Alouds are questions we asked ourselves or issues that arose throughout the SIOP® unit planning process. We thought it would be helpful to include them as a possible resource for your own planning.

We hope that you will use the models we have provided in this chapter as a guide for your own planning and teaching. You may find that some of the lessons will require more or less time than what is allotted here, depending on your students. That is fine; the lessons are designed to show how SIOP® features are integrated into math lessons over the course of a unit of study. We encourage you to adapt them for your own students' needs. Please note that the blackline masters (BLMs) referred to in the unit are found in Appendix C.

SIOP® Planning Flow Chart

Grade Level: 9

Unit Concept: (Big Idea): Transforming Parent Functions

Subject: Math

Approximate Time Involved: 5 days

Lesson Focus 1:
Review of Representing Functions in Multiple Ways

Content Objectives:
SW represent an exponential function in multiple ways.

Language Objectives:
SW define key vocabulary by completing 4-Corners Vocabulary Charts.
SW discuss in small groups how bacteria grow.

Reading/Writing/Discussion Activities:
KWL, 4-Corners Vocabulary, Multiple Representations Graphic Organizer, Think, Pair, Share, Video

Lesson Focus 2:
Comparing and Contrasting Functions

Content Objectives:
SW compare and contrast the attributes and characteristics of linear, quadratic, and exponential functions.
SW identify the parent functions of linear, quadratic, and exponential functions.

Language Objectives: SW express in writing the similarities and differences of linear, quadratic and exponential functions by completing a Function Matrix and sharing their findings in small groups.

(continued)

Reading/Writing/Discussion Activities:
Function Matrix, KWL, Station Activities, 4-Corners Vocabulary, Think, Pair, Share, Multiple Representations Graphic Organizer

Lesson Focus 3:
Translating Parent Functions

Content Objective:
SW investigate transformations of linear, quadratic, and exponential functions.

Language Objective:
SW use words and illustrations to describe the meaning of *transformations* by participating in a Graffiti Write.

Reading/Writing/Discussion Activities:
Graffiti Write, Cupid Shuffle Line Dance, KWL, Graph Yourself on Coordinate Grid

Lesson Focus 4:
Exploring More Transformations of Parent Functions

Content Objective:
SW investigate transformations of linear, quadratic, and exponential functions.

Language Objective:
SW use words, tables, and graphs to explain the effects of transformations on linear, quadratic, and exponential parent functions.

Reading/Writing/Discussion Activities:
Go to Your Corner, 4-Corner Vocabulary, Conga Line, Station Activities, KWL

SIOP® LESSON PLAN, *Grade 9, Days 1–2 Representing Functions in Multiple Ways (Reference Chapter 3, Bacteria Growth and Exponential Functions)*

Class/Subject Area(s): Math
Unit/Theme: Transforming Parent Functions

Grade Level: 9
Lesson Duration: 90 minutes

NCTM Standards

Content	Process
Algebra: • Understand and perform transformations such as arithmetically combining, composing, and inverting commonly used functions, using technology to perform such operations on more complicated symbolic expressions. • Understand relations and functions and select, convert flexibly among, and use various representations for them. • Identify essential quantitative relationships in a situation and determine the class or classes of functions that might model the relationships. **Geometry:** • Understand and represent translations, reflections, rotations, and dilations of objects in the plane by using sketches, coordinates, vectors, function notation, and matrices.	☒ Problem Solving ☒ Reasoning & Proof ☒ Communication ☒ Connections ☒ Representations

Content Objective(s):
SW represent an exponential function in multiple ways.

Language Objective(s):
SW define key vocabulary by completing 4-Corners Vocabulary Charts.
SW discuss in small groups how bacteria grow.

Key Vocabulary:		*Supplementary Materials:*

Content Vocabulary
- Independent variable
- Dependent variable
- Linear function
- Exponential function
- Quadratic function
- Discrete situation
- Continuous situation
- Domain
- Range
- Rate of change
- Function rule
- Exponential growth
- Exponential decay
- Multiple representations

Functional Vocabulary
- Represent
- e-coli bacteria

- Math representations graphic organizer (MRGO) BLM 2
- KWL
- Clay

- Index cards
- Bacteria growth video
- Chart paper
- Examples of linear, quadratic, exponential function graphs

SIOP® Features:

Preparation
- ✗ Adaptation of content
- ✗ Links to background
- ✗ Links to past learning
- ✗ Strategies incorporated

Scaffolding
- ✗ Modeling
- ✗ Guided practice
- ✗ Independent practice
- ✗ Comprehensible input

Grouping Options
- ✗ Whole class
- ✗ Small groups
- ✗ Partners
- ✗ Independent

Integration of Processes
- ✗ Reading
- ✗ Writing
- ✗ Speaking
- ✗ Listening

Application
- ✗ Hands-on
- ✗ Meaningful
- ✗ Linked to objectives
- ✗ Promotes engagement

Assessment
- __ Individual
- ✗ Group
- ✗ Written
- ✗ Oral

Lesson Sequence:

1. To informally assess what students know about linear, quadratic, and exponential functions, show a graph of each function. Provide students with a copy of the KWL chart and have them write what they know about these functions on the K section of the KWL. Also, have them write what they know about transformations. Once all students are done, have a whole group discussion. Construct a KWL on chart paper, and in the K section, scribe what students know about these functions and transformations. Inform students that this unit will focus on investigating the effects transformations have on parent functions of the above-named functions. Instruct students to write in the W section of the KWL what they would like to learn about in this unit. Scribe students' reflections on the KWL chart paper. Prior to starting with transformations of parent functions, tell students they will review essential vocabulary terms, review how to represent a function in multiple ways, and review linear, quadratic, and exponential functions by comparing and contrasting their attributes and characteristics.

2. Introduce the lesson's content and language objectives.

3. Provide a linguistic and nonlinguistic description of the lesson's vocabulary. Allow students to process their understanding of the vocabulary by completing a 4-Corners Vocabulary Chart for each term.

4. To review representing a function in multiple ways, inform students they will investigate the growth of bacteria. Provide each student with an index card. On the index card have students answer the following question: "What do you know about bacteria?" To answer the question, instruct students to use the following sentence starter: "What I know about bacteria is . . ." Once all students are done writing, have

(continued)

SIOP® LESSON PLAN, *Grade 9, Days 1–2 Representing Functions in Multiple Ways (Reference Chapter 3, Bacteria Growth and Exponential Functions)* (continued)

students share their responses by participating in a Think, Pair, Share. After sharing their responses with a partner, select two or three students to state to the whole group what they know about bacteria.

5. In small groups, have students discuss how bacteria grow. Once all groups are done discussing, select two groups to share what they discussed. If students do not know how bacteria grow, build background by showing them a video on bacteria growth. (A video on bacteria growth can be found at the following website: http://www.youtube.com/watch?v=gEwzDydciWc&feature=related.)

6. Tell students they will use clay to model the growth of e-coli bacteria. Explain that an e-coli bacterium splits into two daughter cells every 20 minutes. Instruct students to use a piece of clay to represent one e-coli bacterium. Tell them to pretend 20 minutes have elapsed and the e-coli bacterium has divided into two daughter cells. Have students create the two daughter cells. Call this stage of growth Phase 1. Ask students to create the total number of e-coli bacteria in Phase 2. Phase 3? Phase 4? (For information on the growth rate and generation time of e-coli bacteria visit the following website: http://www.textbookofbacteriology.net/growth.html.)

7. Now that students have represented the growth of e-coli bacteria concretely, have them use BLM 2 Math Representations Graphic Organizer (MRGO) to represent Phase 5 and Phase 6 pictorially. Once students are done drawing Phase 5 and 6 on the graphic organizer, have them organize their data in the table section of the MRGO. The first column will represent the phase number and the third column will represent the number of bacteria. Ask students to determine the number of bacteria starting with Phase 0 and ending with Phase 9. Instruct students to use the middle column to discover and discuss what is happening algorithmically. Have students determine how many bacteria there will be in Phase 30 and Phase X.

8. In the verbal/written section of the MRGO, have students identify the independent variable and dependent variable. Tell students to write a reasonable domain and range and determine whether this problem situation is discrete or continuous. Have students discuss how bacteria grows. Was the rate of change constant? If the answer is no, how did the bacteria grow? Instruct them to write their findings in the verbal/written section.

9. Have students graph the table's data. Remind students to title the graph, label the axes appropriately, and determine reasonable intervals for the axes.

10. In the function rule section, instruct students to write the function rule of this problem situation.

11. At this time, review the attributes and characteristics of exponential functions. Have students determine if the function represents exponential growth or decay. Tell students the parent function of exponential functions is $y = b^x$.

12. As a final wrap-up activity, have students complete the L section of the KWL. Scribe students' responses on the KWL chart paper.

13. Review the lesson's content and language objectives with students.

Reflections:
After teaching the lesson, the teacher reflects on what worked, what did not work, and what revisions, additions, and/or deletions need to be made.

THINK-ALOUDS for SIOP® Lesson Plan, Grade 9, Days 1–2

- How can I help ELs who enter my classroom with little or interrupted schooling? How can I find out if content concepts are appropriate for the age and educational background level of the students? Even though this unit assumes students have some knowledge of linear, quadratic, and exponential functions, lessons 1 and 2 provide a brief review of these functions. The review will help ELs who have little or no understanding of the types of functions studied in ninth grade.

- How will I modify oral discussions for students who are at the initial stages of second language acquisition?

- Refer to Chapter 3 for a full discussion of KWL, Math Representations Graphic Organizer, and 4-Corners Vocabulary.

- ELs are challenged with learning math content and language at the same time. Therefore, setting daily content and language objectives for ELs is extremely important.

PLANNING POINTS for SIOP® Lesson Plan, Grade 9, Days 1–2

- Based on our experience, learning to write language objectives was not an easy task. Like you, we initially felt frustrated and overwhelmed with the idea of having to write both content and language objectives. We knew little about second language acquisition theory, the needs of ELs, and sheltered instruction. However, with sustained SIOP® professional development, practice, and patience, the writing of language objectives became easier.

- The KWL serves as a tool to access prior knowledge. Students should complete the K and W sections in this lesson. In subsequent lessons, students should always add what they learned in the L section.

SIOP® LESSON PLAN *Grade 9, Day 3 Comparing and Contrasting Functions*

Class/Subject Area(s): Math
Unit/Theme: Transforming Parent Functions

Grade Level: 9
Lesson Duration: 60 minutes

NCTM Standards	
Content	*Process*
Algebra:	
• Understand and perform transformations such as arithmetically combining, composing, and inverting commonly used functions, using technology to perform such operations on more-complicated symbolic expressions.	☒ Problem Solving
• Understand relations and functions and select, convert flexibly among, and use various representations for them.	☒ Reasoning & Proof
• Identify essential quantitative relationships in a situation and determine the class or classes of functions that might model the relationships.	☒ Communication
Geometry:	☒ Connections
• Understand and represent translations, reflections, rotations, and dilations of objects in the plane by using sketches, coordinates, vectors, function notation, and matrices.	☒ Representations

Content Objective(s):
SW compare and contrast the attributes and characteristics of linear, quadratic, and exponential functions.

(continued)

SIOP® LESSON PLAN *Grade 9, Day 3 Comparing and Contrasting Functions* (continued)

SW identify the parent functions of linear, quadratic, and exponential functions.

Language Objective(s):
SW express in writing the similarities and differences of linear, quadratic, and exponential functions by completing a Function Matrix and sharing findings in small groups.

Key Vocabulary:			*Supplementary Materials:*	
Content Vocabulary	**Functional Vocabulary**	● KWL		● Tiling Squared Pools Lab Sheet, BLM 21
● Independent variable	● Attributes	● Graphing calculators (optional)		● Fencing Gardens Lab Sheet, BLM 22
● Dependent variable	● Characteristics			● Layers Galore Lab Sheet, BLM 23
● Linear function	● Compare			● Function Matrix BLM 24
● Exponential function	● Contrast			● Color tiles
● Quadratic function	● Similarities			● Paper
● Discrete situation	● Differences			
● Continuous situation				
● Domain				
● Range				
● Rate of change				
● Function rule				
● Exponential growth				
● Exponential decay				
● Multiple representations				

SIOP® Features:

Preparation
- ✗ Adaptation of content
- __ Links to background
- ✗ Links to past learning
- ✗ Strategies incorporated

Scaffolding
- ✗ Modeling
- ✗ Guided practice
- ✗ Independent practice
- ✗ Comprehensible input

Grouping Options
- ✗ Whole class
- ✗ Small groups
- ✗ Partners
- ✗ Independent

Integration of Processes
- ✗ Reading
- ✗ Writing
- ✗ Speaking
- ✗ Listening

Application
- ✗ Hands-on
- ✗ Meaningful
- ✗ Linked to objectives
- ✗ Promotes engagement

Assessment
- __ Individual
- ✗ Group
- __ Written
- ✗ Oral

Lesson Sequence:

1. Give students some quiet time to reflect on what they learned in the previous lesson. Have students participate in a Think, Pair, Share. Select a couple of students to share their reflections. Ask the whole group, "What type of function did the Bacteria Growth activity model have? How did you know? Why was it not linear or quadratic? What are some special attributes of exponential functions?"

2. This lesson's activities will provide students a review of the attributes and characteristics of linear, quadratic, and exponential functions by having them collect data from three problem situations. Students will analyze the data and determine what type of function best represents each problem situation.

3. Introduce content and language objectives.

4. Set up 6 activity stations. Stations 1 and 4: Tiling Squared Pools, BLM 21; Stations 2 and 5: Fencing Gardens, BLM 22; and Stations 3 and 6: Layers Galore, BLM 23. Give students fifteen minutes to perform the activity, collect data, and complete the activity sheet. Once time is up, rotate the groups to the next station until all students have done each activity.

5. In small groups, have students compare and contrast the three activities by completing the Function Matrix, BLM 24. Ask the following questions: What are similarities among the graphs, tables, functions, etc.? How are they different? Which data modeled a linear function? Quadratic? Exponential?

Is Constructing Square Pools quadratic or exponential? Why? Do the problem situations have the same domain and range? How do you know? Are the functions discrete or continuous? How do you know? Identify one spokesperson from each group and have him or her share their findings.

6. Provide direct instruction on the attributes and characteristics of linear, quadratic, and exponential functions and identify corresponding parent functions.

7. Tell students to take out their KWL and write in the L section what they learned today. Add student responses to the KWL chart paper. Close lesson by reviewing content and language objectives.

Reflections:
After teaching the lesson, the teacher reflects on what worked, what did not work, and what revisions, additions, and/or deletion need to be made.

THINK-ALOUDS for SIOP® Lesson Plan, Grade 9, Day 3

- In this lesson, students have several opportunities to participate in whole group discussions. How will I ensure ELs participate in whole group discussions? Idea: After asking a question, I might ask students to think quietly about the answer. Students can then share their answer with a partner and then I can call randomly on students.

- The Function Matrix requires a good deal of writing. How will I help my lower level language students complete the matrix? Lower level language learners might be able to explain the similarities and differences by using gestures or explaining their thoughts to a peer who speaks the EL's native language. I might also be able to work with a small group of ELs as they fill out the matrix.

- Refer to Chapter 3 for a full discussion of KWL and Math Representations Graphic Organizer.

PLANNING POINTS for SIOP® Lesson Plan, Grade 9, Day 3

- Each function has special attributes and characteristics. When providing direct instruction on the attributes and characteristics of functions, consider having EL students take notes on a scaffolded outline (partially completed).

- Center activities can be fun and engaging for high school students. However, it is necessary to be clear in the explanation of academic tasks so that everyone knows his or her responsibility.

SIOP® LESSON PLAN *Grade 9, Day 4 Translating Parent Functions (Reference Chapter 3, Translating Quadratic Functions)*

Class/Subject Area(s): Math
Unit/Theme: Transforming Parent Functions

Grade Level: 9
Lesson Duration: 60 minutes

NCTM Standards

Content	Process
Algebra: - Understand and perform transformations such as arithmetically combining, composing, and inverting commonly used functions, using technology to perform such operations on more complicated symbolic expressions.	☐ Problem Solving

(continued)

SIOP® LESSON PLAN *Grade 9, Day 4 Translating Parent Functions (Reference Chapter 3, Translating Quadratic Functions)* (continued)

- Understand relations and functions and select, convert flexibly among, and use various representations for them.
- Identify essential quantitative relationships in a situation and determine the class or classes of functions that might model the relationships.

Geometry:
- Understand and represent translations, reflections, rotations, and dilations of objects in the plane by using sketches, coordinates, vectors, function notation, and matrices.

☒ Reasoning & Proof

☒ Communication

☒ Connections
☒ Representations

Content Objective(s):
SW investigate transformations of linear, quadratic, and exponential functions.

Language Objective(s):
SW use words and illustrations to describe the meaning of *transformations* by participating in a Graffiti Write.

Key Vocabulary:		Supplementary Materials:

Content Vocabulary
- Transformation
- Translation
- Constant
- Vertical shift
- Horizontal shift

Functional Vocabulary
- Investigate

- Translating Parent Functions Lab Sheet, BLM 25
- 10ft by 10ft coordinate grid
- Masking tape
- Graphing calculator
- Cupid Shuffle Song
- Response boards
- Dry erase markers
- Ordered pairs cards for quadratic parent functions, BLM 26
- Ordered pairs cards for linear parent functions, BLM 27
- Ordered pairs cards for exponential parent functions, BLM 28

SIOP® Features:

Preparation
✗ Adaptation of content
✗ Links to background
✗ Links to past learning
✗ Strategies incorporated

Scaffolding
✗ Modeling
✗ Guided practice
✗ Independent practice
✗ Comprehensible input

Grouping Options
✗ Whole class
✗ Small groups
✗ Partners
✗ Independent

Integration of Processes
✗ Reading
✗ Writing
✗ Speaking
✗ Listening

Application
✗ Hands-on
✗ Meaningful
✗ Linked to objectives
✗ Promotes engagement

Assessment
✗ Individual
✗ Group
✗ Written
__ Oral

Lesson Sequence:

1. Use the KWL chart paper to review what students learned in previous lessons. Inform students they will learn about transforming the parent functions $y = x$, $y = x^2$, and $y = b^x$.

2. Introduce content and language objectives.

3. Activate students' prior knowledge by having them participate in a Graffiti Write: Give students 2 minutes to provide linguistic and nonlinguistic representations of what the word *transformation* means to them. As students are writing/drawing, walk around the room to determine what they know about the word. Once time is up, select a couple of students to share their thoughts. If students do not remember or do not know the mathematical meaning of *transformation,* build background by having them participate in the Cupid Shuffle line dance.

4. Tell students that in the Cupid Shuffle line dance, a person moves 4 steps to the right, 4 steps to the left, kicks 4 times, turns 90 degrees counterclockwise, and starts the pattern all over again. Ask students if they remember from middle school the math terms that describe moving to the right, left, up or down and turning. Tell students *translate* and *rotate* are two forms of mathematical transformation. As a whole group, line dance to the Cupid Shuffle. Reiterate that today's lesson deals with translating parent functions.

5. Review the lesson's key vocabulary by providing a brief linguistic and nonlinguistic explanation of the terms. Once you are done with the explanations, have students complete 4-Corners Vocabulary Charts for each word.

6. Use the Translating Parent Functions Lab Sheet BLM 25 and a graphing calculator to model three examples of translating $y = x$, $y = x^2$, and $y = b^x$. Provide a copy of the Translating Parent Functions Lab Sheet, and have students work in pairs to complete the rest of the worksheet.

7. Have a whole group discussion to compare and contrast the change in size, orientation, and position of the pre-image and image. Ask students: How did the x and y values change in the table? How did the graph change? How did the parent function rule change? When did a function translate horizontally? When did a function translate vertically? Did the translated function change in size? How do you know?

8. To assess students' understanding of translations, construct a 10 feet by 10 feet coordinate plane on the floor. Group students in sets of 5. Provide each group with response boards, markers, and 5 ordered pairs representing one of the parent functions (BLM 26, 27, 28). Explain each group's responsibility will be to graph themselves on the coordinate plane. Once they are graphed on the grid, the teacher will tell them how many units to translate. The rest of the students will write the equation of the translated parent function on their response boards. Model a problem for students.

9. Instruct the first group to graph themselves on the coordinate grid based on their ordered pairs. Once they have graphed the parent function, tell them how many units to translate. After each translation, the students who are sitting write the new function rule on their respond boards. Have students compare their findings. Repeat step 8 until all groups are done.

10. Have students take out their KWL. In the L section, instruct students to write what they learned in this lesson. Add what students learned in the KWL chart paper.

11. Wrap up today's lesson by reviewing the content and language objectives.

Reflections:
After teaching the lesson, the teacher reflects on what worked, what did not work, and what revisions, additions, and/or deletions need to be made.

THINK-ALOUDS for SIOP® Lesson Plan, Grade 9, Day 4

- Think Aloud: For many recent immigrants, a line dance is a new concept. Dancing to the Cupid Shuffle might be fun and engaging for students. However, dancing may be offensive for some ELs due to their cultural or religious beliefs. How can I find out if dancing is offensive for some of my ELs? How can I modify this activity if dancing is offensive to some of them?

- *Translation* is a multiple meaning word. Some ELs might say translation means translating from their native language to English. Provide an explicit explanation of the mathematical meaning of translation. The Cupid Shuffle line dance will help clarify this concept.

PLANNING POINTS for SIOP® Lesson Plan, Grade 9, Day 4

- Refer to Chapter 3 for a full discussion of 4-Corners Vocabulary and Response Boards, and KWL.

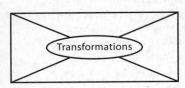

- Graffiti Write is an activity that can be used to access prior knowledge or review content concepts or key vocabulary. In this lesson, Graffiti Write was used to access prior knowledge. In a Graffiti Write, students are grouped in sets of 3–5. The concept or vocabulary term is written in the middle of chart paper. The chart paper is then divided into the number of students per group. Students are given 2–3 minutes to write and draw what they know or learned about the word in the center of the paper. An example is as follows:

SIOP® LESSON PLAN, *Grade 9, Day 5 Exploring More Transformation of Parent Functions*

Class/Subject Area(s): Math
Unit/Theme: Transforming Parent Functions

Grade Level: 9
Lesson Duration: 60 minutes

NCTM Standards

Content	Process
Algebra: • Understand and perform transformations such as arithmetically combining, composing, and inverting commonly used functions, using technology to perform such operations on more complicated symbolic expressions. • Understand relations and functions and select, convert flexibly among, and use various representations for them. • Identify essential quantitative relationships in a situation and determine the class or classes of functions that might model the relationships. **Geometry:** • Understand and represent translations, reflections, rotations, and dilations of objects in the plane by using sketches, coordinates, vectors, function notation, and matrices.	☒ Problem Solving ☒ Reasoning & Proof ☒ Communication ☒ Connections ☒ Representations

Content Objective(s):
SW investigate transformations of linear, quadratic, and exponential functions.

Language Objective(s):
SW use words, tables, and graphs to explain the effects of transformations on linear, quadratic, and exponential parent functions.

Key Vocabulary:		Supplementary Materials:
Content Vocabulary • Transformation • Reflection • Compression • Stretch • Constant • Vertical shift • Horizontal shift	**Functional Vocabulary** • Investigate	• Graphing calculators • Performance Based Assessment (Mosaic Problem) • Go to Your Corner Cards BLM 29 • Multiplying x by $-1 < a < 0$ Lab Sheet, BLM 30 • Multiplying x by $0 < a < 1$ Lab Sheet, BLM 31 • Multiplying x by $a > 1$ Lab Sheet, BLM 32 • Multiplying x by $a < -1$ Lab Sheet, BLM 33 • Multiplying x by -1 Lab Sheet, BLM 34 • Combining Transformations Lab Sheet, BLM 35 • Index cards

Preparation
☒ Adaptation of content
☒ Links to background
☒ Links to past learning
☒ Strategies incorporated

Scaffolding
☒ Modeling
☒ Guided practice
☒ Independent practice
☒ Comprehensible input

Grouping Options
☒ Whole class
☒ Small groups
☒ Partners
☒ Independent

Integration of Processes
☒ Reading
☒ Writing
☒ Speaking
☒ Listening

Application
☒ Hands-on
☒ Meaningful
☒ Linked to objectives
☒ Promotes engagement

Assessment
☒ Individual
☐ Group
☒ Written
☒ Oral

Lesson Sequence:

1. To review translating parent functions, have students participate in the "Go to Your Corner" activity. Post three chart papers around the room and on one chart paper write $y = x$, on another write $y = x^2$, and on the third one write $y = b^x$. Give each student a function card. Their responsibility is to identify their card's parent function and go to the correct corner. Once all students are at their corresponding corners, have them tell their corner partners how their function was translated. Students must use the vocabulary terms they learned to describe translations. Select a couple of students to share how their function was translated.

2. Read the content and language objectives.

3. Introduce the lesson's vocabulary verbally and pictorially. Have students process their understanding of the words by adding them to 4-Corners Vocabulary Charts.

4. Set up 6 activity stations. Station 1: Multiplying x by $a > 1$, BLM 32; Station 2: Multiplying x by $a < -1$, BLM 33; Station 3: Combining Transformations, BLM 35; Station 4: Multiplying x by $0 < a < 1$, BLM 31; Station 5: Multiplying x by $-1 < a < 0$, BLM 30; Station 6: Multiplying x by -1, BLM 24. Each station will need one calculator per student and copies of the corresponding lab sheets. Give students 5 minutes to complete the activity. Once time is up, rotate the groups to the next station until all students have done each activity.

5. Keep students divided in 6 groups. Assign each group an activity station transformation. Each group will be responsible for presenting their findings of their assigned transformation. Group 1 will present Multiplying x by $a>1$; Group 2: Multiplying x by $a < -1$; Group 3: Combining Transformations; Group 4: Multiplying x by $0 < a < 1$; Group 5: Multiplying x by $-1 < a < 0$; Group 6: Multiplying x by -1. Students must use the unit's/lesson's vocabulary terms.

6. Use a performance-based assessment problem to assess students' understanding. Consider using The Mosaics problem from Dana Center's **Algebra I Assessments**. This problem can be found at: http://www.utdanacenter.org/mathtoolkit/downloads/alg1assess/alg1_mosaics.pdf. In addition to the Mosaic problem questions, ask students to identify The Mosaics parent function, describe The Mosaic function's transformations, and discuss what each transformation represents within the problem situation.

7. As a final wrap-up, tell students to think about how parent functions differ from transformed functions. Instruct them to write their thoughts on an index card. They can use the sentence starter: "One difference between parent functions and transformed functions is . . ." Have them share their reflections by participating in a "Conga Line" activity. Tell students to take out their KWL and write in the L section what they learned in this lesson. Scribe their thoughts on the KWL chart paper. Review the L section with students.

8. Close the lesson by reviewing content and language objectives.

Reflections:
After teaching the lesson, the teacher reflects on what worked, what did not work, and what revisions, additions, and/or deletions need to be made.

THINK-ALOUDS for SIOP® Lesson Plan, Grade 9, Day 5

- Students with low levels of English proficiency often have difficulty articulating their thoughts. Therefore, how can a student with limited English proficiency explain the effects transformations cause on parent functions? I should remember that students do not need to rely solely on language to understand and explain this. Other than words, students can use tables, graphs, and equations to demonstrate understanding of the effects of transformations. This lesson's language objective was carefully chosen with this in mind. All students benefit from using words, tables, and graphs to explain the effects of transformations. However, it is a necessary instructional accommodation for students who lack English proficiency.

PLANNING POINT for SIOP® Lesson Plan, Grade 9, Day 5

- Academic vocabulary must be introduced, processed, used, and reviewed. Provide students with opportunities to play games like *Jeopardy*, *Pictionary*, and *Charades* with the unit's words. These games captivate students' interest and elevate levels of student engagement. The Internet is a great place to find teacher-made content/ vocabulary games. Visit the following website to find an array of content-specific games: http://www.hardin.k12.ky.us/res_techn/countyjeopardygames.htm
- Refer to Chapter 4 for a full discussion of Conga Line.

Concluding Thoughts

In this chapter we presented lessons in a unit as models of effective SIOP® mathematics lessons for high school students. Also included is insight into the type of thinking process involved in planning SIOP® lessons. You'll notice that the lessons were designed with attention given to language objectives in every lesson. For English learners to make progress in mathematics—and all subjects—they must be given opportunities to use English for academic purposes. It is impossible to separate out the process of understanding and learning mathematical concepts, operations, processes, and procedures from their language requirements. Notice as well that the language objectives have been designed to build over the course of a unit, so that we advance students' knowledge and skills in the language of mathematics. Finally, we hope the lessons and units illustrated clearly to you how to implement the features of the SIOP® Model in everyday lessons, consistently and cohesively throughout a unit of study.

Pulling It All Together

As we planned this final chapter, we decided it might be helpful if we shared some of the things we have learned in the process of writing this book, related to collaborations with our content contributors and our understandings of lesson planning and teaching using the SIOP® Model. We also asked our contributors to share what they have learned, and we have included their thoughts and insights in the second half of this chapter.

What We Have Learned

One important finding for all of us is the confirmation that becoming an effective SIOP® teacher is a process that takes time, reflectivity, practice, and commitment. Unlike many of the educational initiatives that we have all been involved in during our careers, the SIOP® Model is not about tweaking our teaching a little here and adding a little something there, while expecting immediate results in our students' academic achievement. Instead, the SIOP® Model is about purposeful planning, consistent attention to teaching the academic language and content of your discipline, and maintaining the belief that all students, including English learners, can reach high academic standards while developing their English proficiency.

We also have become even more aware that good teaching is about attention to detail. For example, as we were reading the lesson and unit plans created by our contributors, occasionally we had to call or email and ask questions about the purpose of a particular handout, the steps to a process, or the name of an activity. This made us realize how important it is to use precise language with English learners (and all learners), both in our speech and in the materials we prepare for them. Consistent labeling for classroom routines, procedures, and activities reduces ambiguity and confusion, and serves as additional scaffolding for ELs.

Another similar insight concerns the role of teachers as content experts. You know where you're going, and what needs to be taught, learned, and assessed. Students, including English learners, don't have this insider information, and sometimes they're academically lost because they don't know what is expected of them. Obviously, this is a primary function of content and language objectives: to point the way and assist students in knowing what to expect. But, when we make assumptions about what students know and can do, we may be basing those assumptions on what *we* know and what *we* can do. Once again, being precise in your use of terms, and carefully explaining and modeling processes and procedures related to the content you're teaching will assist your English learners in becoming more successful in learning your content.

We also sharpened our skills in designing SIOP® lessons and units. Even though our contributors created and wrote the lessons, we became even more aware of how challenging it is to write detailed SIOP® lesson and unit plans. Teachers new to the SIOP® Model often balk at the time and work it will take to plan lessons, yet as we have mentioned previously, with practice, the amount of time and effort is diminished. The end goal needs to be kept in sight: the academic and language proficiency benefits for the English learners who will be productive members of our society in the future.

Finally, we have learned from our contributors that experienced, knowledgeable, successful teachers who are well-versed in the SIOP® Model continue to grow and learn through the process of carefully planning and teaching effective and appropriate lessons for English learners. In the section that follows, you will hear their voices about what they learned during the process of working together to help write this book.

What Our Contributors Have Learned

It was a pleasure collaborating with our math contributors and they were generous in taking time to reflect on their own experience. The comments that follow reflect their personal insights about working with others, the lesson and unit planning process and teaching using the SIOP® Model. We hope you will find their comments as useful as we did.

Araceli Avila—Mathematics

I learned that collaborating with team members is so important. It was interesting to see how all content teams interpreted the guidelines in a different manner. The end in mind was the same, but how we got there was different. This experience reminded me that it is more important to get to our final destination than to spend our energy focusing on how we get there. (Editor's Note: There were also content terms in the areas of English-Language Arts, History-Social Studies and Sciences.)

The SIOP® Unit Flow Chart gave me direction and focus. I was able to see a progression of the content objectives at a glance. It also made the writing of the lessons so much easier.

As educators, sometimes we like experiences to be canned because it makes our jobs easier. However, when we provide students with open-ended tasks, they are able to apply their own strategies and come up with great outcomes.

I made sure to include as many language opportunities in my lessons as I could. I found it easier to write the language objective once the lesson was finished. (*Authors' note*: This was interesting to us because we generally recommend starting with the objectives first when planning a lesson. However, Araceli points out that when her focus was on incorporating language learning activities in her math lessons, she could then go back and craft the language objectives for the students.)

Coming up with higher order thinking questions (consistently) was difficult for me. Therefore, I know that I need additional professional development in this area. Self-awareness is the key to change. This project helped me identify some of my own weaknesses. However, self-awareness in itself is nothing. Changing what needs to be changed is the goal.

I loved participating in the project because the SIOP® Model makes me reflect on my own teaching. It also allows me the opportunity to become a better educator and in turn become a better professional developer.

Melissa Castillo—Mathematics

As I reflect on the impact that the SIOP® Model has had on my ability to be an effective educator, I am tremendously grateful for the blessings of having been trained in the model, and now having the opportunity to train the SIOP® Model myself (with Pearson Education). As overstated as it might sound, the SIOP® Model has changed my professional life. I now train teachers across the country on how to best address the needs of English learners, and I am able to do so in a way that not only introduces them to the approaches necessary for their success, but gives them the opportunity to practice them as well, so that they truly experience and understand the impact they can have on students. When teaching, I feel I am now the teacher that children need to achieve academically and develop true proficiency in English.

As far as my participation in the writing of the Math Content Book, I have truly been humbled by the amount of work, time, and energy it takes for teachers to prepare effective SIOP® lessons. I do, however, believe that the investment made is well worth it. During this process, I found it was interesting to consider once again what quality instruction is, and how to include it in a SIOP® lesson. I believed it was obvious and quite clear. However, as my work was reviewed and I was given feedback and asked questions from peers and experts on the SIOP® Model, I quickly realized that what was obvious to me was not always clear to others. The question then became: Can the lessons be comprehensible for learners? What I hope those of you who are reading this will understand is that the opportunity to have others help you reflect is a valuable one. Planning and delivering SIOP® instruction is not an individual process, but a collaborative one. I can only hope that my involvement in the ongoing movement of the SIOP® Model will be long lived!

Final Thoughts

As you read the thoughts and insights of our contributors, you may have noticed how they mentioned the importance of collaboration. We couldn't agree more. For over fifteen years, we have collaborated about the SIOP® Model, with each other and with educators throughout the country (and now throughout the world). We are convinced that these collaborations have resulted in a comprehensive model of instruction for English learners that is not only empirically validated, but is appropriate, essential, and *doable* for teachers, despite the time and effort it takes to create, write, and then teach SIOP lessons.

We hope this book, and the others in the SIOP® series, help you pull it all together as you create your SIOP® mathematics lesson plans. We also hope that as you become a successful SIOP® teacher, you will find rewards in your students' academic and language growth, and more confidence in how to effectively teach mathematics to your English learners.

appendix a: SIOP® Protocol and Component Overview

(Echevarria, Vogt, & Short, 2000; 2004; 2008)

The SIOP® Model was designed to help teachers systematically, consistently, and concurrently teach grade-level academic content and academic language to English learners (ELs). Teachers have found it effective with both ELs and native-English speaking students who are still developing academic literacy. The model consists of eight components and 30 features. The following brief overview is offered to remind you of the preparation and actions teachers should undertake in order to deliver effective SIOP® instruction.

1. Lesson Preparation

The focus for each SIOP® math lesson is the content and language objectives. We suggest that the objectives be linked to curriculum standards and the academic language students need for success in math. Your goal is to help students gain important experience with key grade-level content and skills as they progress toward fluency in English. Hopefully, you now post and discuss the objectives with students each day, even if one period continues a lesson from a previous day, so that students know what they are expected to learn and/or be able to do by the end of that lesson. When you provide a road map at the start of each lesson, students focus on what is important and take an active part in their learning process.

The Lesson Preparation component also advocates for supplementary materials (e.g., visuals, multimedia, adapted or bilingual texts, study guides) because grade-level texts are often difficult for many English learners to comprehend. Graphics or illustrations may be used to make content meaningful—the final feature of the component. It is important to remember that meaningful activities provide access to the key concepts in your math lessons; this is much more important than just providing "fun" activities that students readily enjoy. Certainly, "fun" is good, but "meaningful" and "effective" are better. You will also want to plan tasks and projects for students so they have structured opportunities for oral interaction throughout your lessons.

2. Building Background

In SIOP® lessons, you are expected to connect new concepts with students' personal experiences and past learning. As you prepare ELs for math lessons, you may at times have to build background knowledge because many English learners either have not been studying in U.S. schools or are unfamiliar with American culture. At other times, you may need to activate your students' prior knowledge in order to find out what they already know, to identify misinformation, or to discover when you need to fill in gaps.

The SIOP® Model places importance on building a broad vocabulary base for students. We need to increase vocabulary instruction across the curriculum so our students will become effective readers, writers, speakers, and listeners—and mathmeticians. As a math teacher, you already explicitly teach key vocabulary. Go even further for your ELs by helping them develop word learning strategies such as using context clues, word parts (e.g., affixes), visual aids (e.g., illustrations), and cognates (a word related in meaning and form to a word in another language). Then be sure to design lesson activities that give students multiple opportunities to use new mathematics vocabulary both orally and in writing, such as those found in this book and in *99 Ideas and Activities for Teaching with the SIOP® Model* (Vogt & Echevarria, 2008). In order to move words from receptive knowledge to expressive use, vocabulary needs reinforcement through different learning modes.

3. Comprehensible Input

If you present information in a way that students cannot understand, such as an explanation that is spoken too rapidly, or texts that are far above students' reading levels with no visuals or graphic organizers to assist them, many students—including English learners—will be unable to learn the necessary content. Instead, modify "traditional" instruction with a variety of ESL methods and SIOP® techniques so your students can comprehend the lesson's key concepts. These techniques include, among others:

- Teacher talk appropriate to student proficiency levels (e.g., simple sentences, slower speech)
- Demonstrations and modeling (e.g., modeling how to complete a task or problem)
- Gestures, pantomime, and movement
- Role-plays, improvisation, and simulations
- Visuals, such as pictures, real objects, illustrations, charts, and graphic organizers
- Restatement, paraphrasing, repetition, and written records of key points on the board, transparencies, or chart paper
- Previews and reviews of important information (perhaps in the native language, if possible and as appropriate)
- Hands-on, experiential, and discovery activities

Remember, too, that academic tasks must be explained clearly, both orally and in writing for students. You cannot assume English learners know how to do an assignment because it is a regular routine for the rest of your students. Talk through the procedures and use models and examples of good products and appropriate participation, so students know the steps they should take and can envision the desired result.

When you are dealing with complicated and abstract concepts, it can be particularly difficult to convey information to less proficient students. You can boost the comprehensibility of what you're teaching through native language support, if possible. Supplementary materials (e.g., adapted texts or CDs) in a student's primary language may be used to introduce a new topic, and native language tutoring (if available) can help students check their understanding.

4. Strategies

This component addresses student learning strategies, teacher-scaffolded instruction, and higher-order thinking skills. By explicitly teaching cognitive and metacognitive learning strategies, you help equip students for academic learning both inside and outside the SIOP® classroom. You should capitalize on the cognitive and metacognitive strategies students already use in their first language because those will transfer to the new language.

As a SIOP® teacher, you must frequently scaffold instruction so students can be successful with their academic tasks. You want to support their efforts at their current performance level, but also move them to a higher level of understanding and accomplishment. When students master a skill or task, you remove the supports you provided and add new ones for the next level. Your goal, of course, is for English learners to be able to work independently. They often achieve this independence one step at a time.

You need to ask your ELs a range of questions, some of which should require critical thinking. It is easy to ask simple, factual questions, and sometimes we fall into that trap with beginning English speakers. We must go beyond questions that can be answered with a one- or two-word response, and instead, ask questions and create projects or tasks that require students to think more critically and apply their language skills in a more extended way. Remember this important adage: "Just because ELs don't speak English proficiently doesn't mean they can't *think*."

5. Interaction

We know that students learn through interaction with one another and with their teachers. They need oral language practice to help develop and deepen their content knowledge and support their second language skills. Clearly, you are the main role model for appropriate English usage, word choice, intonation, fluency, and the like, but do not discount the value of student–student interaction. In pairs and small groups, English learners practice new language structures and vocabulary that you have taught as well as important language functions, such as asking for clarification, confirming interpretations, elaborating on one's own or another's idea, and evaluating opinions.

Don't forget that sometimes the interaction patterns expected in an American classroom differ from students' cultural norms and prior schooling experiences. You will want to be sensitive to sociocultural differences and work with students to help them become competent in the culture you have established in your classroom, while respecting their values.

6. Practice & Application

Practice and application of new material is essential for all learners. Our research on the SIOP® Model found that lessons with hands-on, visual, and other kinesthetic tasks benefit ELs because students practice the language and content knowledge through multiple modalities. As a SIOP® teacher, you want to make sure your lessons include a variety of activities that encourage students to apply both the mathematics content and the English language skills they are learning.

7. Lesson Delivery

If you have delivered a successful SIOP® lesson, that means that the planning you did worked—the content and language objectives were met, the pacing was appropriate, and the students had a high level of engagement. We know that lesson preparation is crucial to effective delivery, but so are classroom management skills. We encourage you to set routines, make sure students know the lesson objectives so they can stay on track, and introduce (and revisit) meaningful activities that appeal to students. Don't waste time, but be mindful of student understanding so that you don't move a lesson too swiftly for students to grasp the key information.

8. Review & Assessment

Each SIOP® lesson needs time for review and assessment. You will do your English learners a disservice if you spend the last five minutes teaching a new concept rather than reviewing what they have learned so far. Revisit key vocabulary and concepts with your students to wrap up each lesson. Check on student comprehension frequently throughout the lesson period so you know whether additional explanations or reteaching are needed. When you assess students, be sure to provide multiple measures for students to demonstrate their understanding of the content. Assessments should look at the range of language and content development, including measures of vocabulary, comprehension skills, and content concepts.

WHY IS THE SIOP® MODEL NEEDED NOW?

We all are aware of the changing demographics in our U.S. school systems. English learners are the fastest growing subgroup of students and have been for the past two decades. According to the U.S. Department of Education in 2006, English learners numbered 5.4 million in U.S. elementary and secondary schools, about 12% of the student population, and they are expected to comprise about 25% of that population by 2025. In several states, this percentage has already been exceeded. The educational reform movement, and the No Child Left Behind (NCLB) Act in particular, has had a direct impact on English learners. States have implemented standards-based instruction and high-stakes testing, but in many content classes, little or no accommodation is made for the specific language development needs of English learners; this raises a significant barrier to ELs' success because they are expected to achieve high academic standards in English. In many states, ELs are required to pass end-of-grade tests in order to be promoted and/or exit exams in order to graduate.

Unfortunately, teacher development has not kept pace with the EL growth rate. Far too few teachers receive an undergraduate education that includes coursework in English as a second language (ESL) methodologies, which can be applied in content classes through sheltered instruction, and in second language acquisition theory, which can help teachers understand what students should be able to accomplish in a second language according to their proficiency levels, prior schooling, and sociocultural backgrounds. At the end of 2008, only four states—Arizona, California, Florida, and New York—required some undergraduate coursework in these areas for all teacher candidates.

Some teachers receive inservice training in working with ELs from their schools or districts, but it is rarely sufficient for the task they confront. Teachers are expected to teach ELs the new language, English, so the ELs can attain a high degree of proficiency, and in addition, instruct them in all the topics of the different grade-level content courses (more often than not taught in English). A survey conducted by Zehler and colleagues (2003) in 2002 found that approximately 43% of elementary and secondary teachers had ELs in their classrooms, yet only 11% were certified in bilingual education and only 18% in English as a second language. In the five years prior to the survey, teachers who worked with three or more ELs had received on average four hours of inservice training in how to serve them—hardly enough to reach a satisfactory level of confidence and competence.

Even teachers who have received university preparation in teaching English learners report limited opportunities for additional professional development. In a recent survey that sampled teachers in 22 small, medium, and large districts in California, the researchers found that during the previous five years, "forty-three percent of teachers with 50 percent or more English learners in their classrooms had received no more than one inservice that focused on the instruction of English learners" (Gandara, Maxwell-Jolly & Driscoll, 2005, p. 13). Fifty percent of the teachers with somewhat fewer students (26%–50% English learners in their classes) had received either no such inservice or only one. The result of this paucity of professional development is that ELs sit in classes with teachers and other staff who lack expertise in second language acquisition, multicultural awareness, and effective, research-based classroom practices.

It is not surprising, then, that ELs have experienced persistent underachievement on high-stakes tests and other accountability measures. On nearly every state and national assessment, ELs lag behind their native-English speaking peers and demonstrate significant achievement gaps (Kindler, 2002; Kober, et al., 2006; Lee, Grigg & Dion, 2007; Lee, Grigg & Donahue, 2007). In addition to having underqualified teachers, ELs are also more likely to be enrolled in poor, majority-minority schools that have fewer resources and teachers with less experience and fewer credentials than those serving English-proficient students (Cosentino de Cohen, Deterding & Clewell, 2005).

Lower performance on assessments is also the result of education policy. Although research has shown that it takes several years of instruction to become proficient in English (four to nine years, depending on a student's literacy level in the native language and prior schooling) (Collier, 1987; Cummins, 2006; Genesee, Lindholm-Leary, Saunders, & Christian, 2006), current NCLB policy forces schools to test ELs in reading after one year of U.S. schooling in grades 3–8 and one grade in high school. English learners are supposed to take the tests in mathematics and science from the start. Adding to the disconnect between research and policy is the fact that these tests have been designed for native English speakers, rendering them neither valid nor reliable for ELs (AERA, APA, & NCME, 2000). By definition, an English learner is **not** proficient in English; as most of these state assessments are in English, the majority of ELs score poorly on them and are unable to demonstrate their real level of understanding of the subject matter.

Even though it is hard to turn around education policy, teachers, schools, districts, and universities do have opportunities to enact changes in professional development and program design. With this book we hope to help English-language arts teachers grow professionally and develop appropriate skills for working with English learners. There are many approaches and numerous combinations of techniques that can be applied to the delivery of sheltered content instruction. Currently, however, the SIOP® Model is the only scientifically validated model of sheltered instruction for English learners, and it has a growing research base (Center for Applied Linguistics, 2007; Echevarria, Richards, Canges & Francis, 2009; Echevarria & Short, in press; Echevarria, Short & Powers, 2006; Short & Richards, 2008). The SIOP® Model is distinct from other approaches in that it offers a field-tested protocol for systematic lesson planning, delivery, and assessment, making its application for teaching English learners transparent for both preservice candidates preparing to be teachers and practicing teachers engaged in staff development. Further, it provides a framework for organizing the instructional practices essential for sound sheltered content instruction.

THE SHELTERED INSTRUCTION OBSERVATION PROTOCOL (SIOP) MODEL

1) **Lesson Preparation**
    1. **Content objectives** clearly defined, displayed and reviewed with students
    2. **Language objectives** clearly defined, displayed and reviewed with students
    3. **Content concepts** appropriate for age and educational background level of students
    4. **Supplementary materials** used to a high degree, making the lesson clear and meaningful (e.g., computer programs, graphs, models, visuals)
    5. **Adaptation of content** (e.g., text, assignment) to all levels of student proficiency
    6. **Meaningful activities** that integrate lesson concepts (e.g., interviews, letter writing, simulations, models) with language practice opportunities for reading, writing, listening, and/or speaking

2) **Building Background**
    7. **Concepts explicitly linked** to students' background experiences
    8. **Links explicitly made** between past learning and new concepts

9. **Key vocabulary emphasized** (e.g., introduced, written, repeated, and highlighted for students to see)

3) **Comprehensible Input**

10. **Speech** appropriate for students' proficiency levels (e.g., slower rate, enunciation, and simple sentence structure for beginners)

11. **Clear explanation** of academic tasks

12. **A variety of techniques** used to make content concepts clear (e.g., modeling, visuals, hands-on activities, demonstrations, gestures, body language)

4) **Strategies**

13. Ample opportunities provided for students to use **learning strategies**

14. **Scaffolding techniques** consistently used, assisting and supporting student understanding (e.g., think-alouds)

15. A variety of **questions or tasks that promote higher-order thinking skills** (e.g., literal, analytical, and interpretive questions)

5) **Interaction**

16. Frequent opportunities for **interaction** and discussion between teacher / student and among students, which encourage elaborated responses about lesson concepts

17. **Grouping configurations** support language and content objectives of the lesson

18. Sufficient **wait time for student responses** consistently provided

19. Ample opportunities for students to **clarify key concepts in Ll** as needed with aide, peer, or L1 text

6) **Practice and Application**

20. **Hands-on materials and / or manipulatives** provided for students to practice using new content knowledge

21. Activities provided for students to **apply content and language knowledge** in the classroom

22. Activities integrate all **language skills** (i.e., reading, writing, listening, and speaking)

7) **Lesson Delivery**

23. **Content objectives** clearly supported by lesson delivery

24. **Language objectives** clearly supported by lesson delivery

25. **Students engaged** approximately 90% to 100% of the period

26. **Pacing** of the lesson appropriate to students' ability levels

8) **Review and Assessment**

27. Comprehensive **review of key vocabulary**

28. Comprehensive **review of key content concepts**

29. Regular **feedback** provided to students on their output (e.g., language, content, work)

30. **Assessment of student comprehension and learning** of all lesson objectives (e.g., spot checking, group response) throughout the lesson

Academic Math Vocabulary Based on NCTM Content & Process Standards Grade Band: K–2

appendix b: Academic Math Vocabulary Based on NCTM Content & Process Standards Grade Band: K–2

Number and Operations	Algebra	Geometry	Measurement	Data Analysis & Probability	Process Standards
Addition	Decreasing pattern	Above	Area	Bar type graph	Guess and check
Cardinal number	Increasing pattern	Behind	Calendar	Chance	Prediction
Coin	Numeric pattern	Below	Clock	Graph	
Difference	Pattern	Between	Day	Lists	
Greater than	Pattern extension	Circle	Distance	Outcome	
Grouping	Shape pattern	Corner	Estimate answer	Picture graph	
Less than	Sound pattern	Direction	Foot		
Money		In front	Height		
Number		Inside	Hour		
Numeral		Left	Inch		
Ordinal number		Location	Length		
Set		Model	Measuring cup		
Subtraction		Near	Minute		
Sum		Shape combination	Pound		
Whole number		Shape division	Second		
Zero		Similarity	Size		
		Under	Standard measure of weight		
			Standard measures of time		
			Table		
			Temperature		
			Temperature Estimation		
			Time interval		
			Volume		
			Week		
			Year		

Source: Developed by Araceli Avila

Academic Math Vocabulary Based on NCTM Content & Process Standards Grade Band: 3–5

Academic Math Vocabulary Based on NCTM Content & Process Standards Grade Band: 3–5

Number and Operations	Algebra	Geometry	Measurement	Data Analysis & Probability	Process Standards
Addend	Constant	2-dimensional decomposition	Angle measurement tool	Bar graph	Diagram
Addition algorithm	Function	2-dimensional shape	Area	Certainly	Invalid argument
Associative property	Geometric pattern	2-dimensional shape combination	Capacity	Cluster	Irrelevant information in a problem
Basic number combination	Geometric pattern extension	2-dimensional shape slide	Centimeter	Data	Process of elimination
Common denominator	Growing pattern	2-dimensional shape turn	Circumference	Data cluster	Proof
Common factor	Linear pattern addition	2-dimensional space	Conservation of area	Data collection method	Relevant information in a problem
Commutative property	Number sentence	3-dimensional shape	Different size units	Event likelihood	Restate a problem
Decimal	Open sentence	3-dimensional shape combination	Elapsed time	Extreme value	Symbolic representation
Decimal addition	Path	Acute angle	English System of Measurement	Histogram	Trial & error
Decimal division	Pattern addition	Angle	Estimation of height	Improbability	Valid argument
Decimal estimation	Pattern subtraction	Angle unit	Estimation of length	Investigation	Venn diagram
Decimal multiplication	Repeating pattern	Classes of triangles	Estimation of width	Line graph	Verbal representation of a problem
Decimal subtraction	Shrinking pattern	Corresponding angles	Gram	Mean	Verification
Distributive property		Corresponding sides	Mass	Measures of central tendency	
Dividend		Cube	Measurement	Median	
Divisibility		Cylinder	Measures of height	Mode	
Division		Equilateral triangle	Measures of length	Pie chart	
Equation		Faces of a shape	Measures of width	Probability	
Equivalent forms		Flip transformation	Meter	Sample	

Source: Developed by Araceli Avila

	Horizontal axis	Metric system	Studies
Equivalent fractions			
Equivalent representation	Intersection of shapes	Parallelogram formula	Survey
Estimation	Isosceles triangle	Perimeter	Tallies
Estimation of fractions	Midpoint	Rectangle formula	Variability
Even numbers	Number of faces	Ruler	
Expanded notation	Obtuse angle	Same size units	
Factors	Parallel lines	Standard vs. nonstandard units	
Fraction	Parallelogram	Surface area	
Fraction addition	Perpendicular lines	Time zone	
Fraction division	Prism	Triangle formula	
Fraction multiplication	Pyramid	Unit conversion	
Fraction subtraction	Rectangular prism	Unit differences	
Fractions of different size	Relative distance	Volume	
Front-end digits	Relative size	Volume of irregular shapes	
Front-end estimation	Rhombus	Volume of rectangular solids	
Front-end estimation	Right angle		
Greatest common factor	Rotation		
Identity property	Scale		
Improper fraction	Shape similarity		
Inequality	Shape transformation		
Inequality solution	Sphere		
Least common multiple	Vertical axis		
Mixed numbers			
Multiple			

(continued)

Academic Math Vocabulary Based on NCTM Content & Process Standards Grade Band: 3–5 (continued)

Number and Operations	Algebra	Geometry	Measurement	Data Analysis & Probability	Process Standards
Multiplication					
Negative number					
Number pairs					
Number triplet					
Odd number					
Order of operations					
Part to whole					
Percent					
Positive number					
Prime factorization					
Prime number					
Product					
Quotient					
Reduced form					
Relative magnitude					
Relative magnitude of fractions					
Remainder					
Reversing order of operations					
Rounding					
Subset					
Subtraction algorithm					
Truncation					
Unlike denominators					

Academic Math Vocabulary Based on NCTM Content & Process Standards Grade Band: 6–8

Academic Math Vocabulary Based on NCTM Content & Process Standards Grade Band: 6–8

Number and Operations	Algebra	Geometry	Measurement	Data Analysis & Probability	Process Standards
Addition of fractions	Algebraic expression	3-dimensional shape	Area of irregular shapes	Area model	Algebraic representation
Algebraic expression expansion	Algebraic step function	3-dimensional shape cross section	Benchmarking	Biased sample	Complex problem
Array	Approximate lines	Alternate interior angle	Circle formula	Box & whisker plot	Conjecture
Base 10	Combining like terms	Angle bisector	Circumference formula	Certainty of conclusions	Counter example
Base 60	Constant difference	Axis of symmetry	Cubic unit	Complementary event	Counting procedure
Composite number	Constant rate of change	Blueprint	Grid	Data display error	Deductive argument
Convert large number to small number	Constant ratio	Complementary angle	Linear units	Data extreme	Deductive prediction
Cube number	Distance formula	Congruence	Perimeter formula	Data gap	Graphic representation of function
Cube root	Equal ratio	Conjecture	Precision of measurement	Data set	Inductive reasoning
Exponent	Equation systems	Coordinate geometry	Range of estimation	Dispersion	Input/output table
Exponential notation	Formula for missing values	Coordinate plane	Reliability	Experiment	Logic NONE
Integer	Growth rate	Coordinate system	Significant digits	Fair chance	Logic ALL
Multiplication algorithm	Intercept	Defining properties of shapes/figures	Square units	Frequency	Logic AND
Nondecimal numeration system	Iterative sequence	Dilation	Straight edge & compass	Frequency distribution	Logic IF/THEN
Number property	Linear arithmetic sequence	Enlarging transformation	Thermometer	Large sample	Logic NOT
Number system	Linear equation	Intersecting lines	Trapezoid formula	Limited sample	Logic OR
Number theory	Linear geometric sequence	Irregular polygon	Unit size	Mutually exclusive events	Logic SOME

(continued)

Source: Developed by Araceli Avila

Academic Math Vocabulary Based on NCTM Content & Process Standards Grade Band: 6–8 (continued)

Number and Operations	Algebra	Geometry	Measurement	Data Analysis & Probability	Process Standards
Odds	Mathematical expression	Line symmetry	Volume formula	Nominal data	Multiple problem solving strategies
Overestimation	Maximum	Ordered pairs	Volume of cylinder	Outliers	Multiple strategies for proof
Prime factor	Networks	Parallel figures	Volume of prism	Random number	Nonroutine vs. routine problems
Rational number	Nonlinear equation	Pattern division	Volume of pyramid	Random sample	Pictorial representation
Reference set	Nonlinear function	Perpendicular bisector		Random variable	Problem area
Relatively prime	Pattern multiplication	Perspective		Range	Problem formulation
Roman numeral	Pattern recognition	Planar cross section		Relative frequency	Problem space
Root	Percents above 100	Plane		Sample selection techniques	Solution algorithm
Scientific notation	Percents below 1	Plane figure		Sample space	Solution probabilities
Square number	Place holder	Polygon		Sampling error	Table representation of functions
Square root	Proportion	Projection		Scale	Work backwards
Underestimation	Proportional gain	Quadrilateral		Scatter plot	Written representation
	Quadratic equation	Reflection transformation		Spreadsheet	
	Rate	Regular coordinates		Stem & leaf plot	
	Rate change	Rotation symmetry		Theoretical probabilities	
	Recursive sequence	Scale drawing		Tree diagram model	
	Sequence	Scale map			
	Similar proportions	Scale transformation			
	Simplification	Shrinking transformation			

Slope	Similarity vs. congruence				
Slope intercept formula	Slide transformation				
Substitution for unknowns	Solid figure				
Unknown	Supplementary angle				
Variable	Tessellation				
Variable change	Tetrahedron				
	Vertex				

Academic Math Vocabulary Based on NCTM Content & Process Standards Grade Band: 9–12

Academic Math Vocabulary Based on NCTM Content & Process Standards Grade Band: 9–12

Number and Operations	Algebra	Geometry	Measurement	Data Analysis & Probability	Process Standards
Absolute value	Absolute function	Angle of depression	Absolute error	Bivariate data	Derivation
Add rational expressions	Acceleration	Arc	Decibel	Bivariate data transformation	Formal mathematical induction
Addition counting procedure	Algebraic function	Cartesian coordinates	Density	Bivariate distribution	Mathematical theories
Base e	Area under curve	Central angle	Direct measure	Categorical data	Monitor progress of a problem
Binary system	Asymptote of function	Chord	Indirect measure	Central limit theorem	Nature of deduction
Complex number	Circular function	Circle without center	Limit	Combination	Smallest set of rules
Compound interest	Classes of functions	Cosine	Parameter	Compound event	Strategy efficiency
Conjugate complex number	Continuity	Dilation of objects in a plane	Parameter estimate	Conditional probability	Strategy generation technique
Divide radical expressions	Curve fitting	Isometry	Precision of estimation	Confidence interval	
Exponent	Curve fitting median method	Line segment	Protractor	Continuous probability distribution	
Fraction inversion	Direct function	Line segment congruence	Relative error	Control group	
Imaginary number	Domain of function	Line segment similarity	Richter scale	Correlation	
Interest	Equivalent forms of equations	Line through point not on a line	Successive approximations	Critical paths method	
Irrational number	Equivalent forms of inequalities	Pi	Surface area cone	Dependent events	
Matrix addition	Exponential function	Point of tangency	Surface area cylinder	Discrete probability	
Matrix division	Fibonacci sequence	Polar coordinates	Surface area sphere	Discrete probability distribution	
Matrix inversion	Force	Postulate	Unit analysis	Empirical verification	

Source: Developed by Araceli Avila

Matrix multiplication	Formal mathematical induction	Proof graph	Upper/lower bounds	Expected value
Matrix subtraction	Function composition	Pythagorean theorem		Experimental design
Multiply radical expressions	Function notation	Radius		Experimental probability
Natural log	Geometric function	Reflection in plane		Factorial
Natural number	Global/local behavior	Reflection in space		Factorial notation
Negative exponents	Inflection	Right triangle geometry		Finite graph
Negative exponents	Inverse function	Rotation in plane		Independent events
Number subsystems	Law of large numbers	Similar figures		Independent trials
Polynomial addition	Line equation	Sine		Law of probability
Polynomial division	Linear	Synthetic geometry		Monte Carlo simulation
Polynomial multiplication	Log function	Tangent		Normal curve
Polynomial subtraction	Logarithm	Theorem direct proof		Parallel box plot
Powers	Logarithmic function	Theorem indirect proof		Permutation
Real numbers	Matrix	Transversal		Population
Reciprocal	Matrix equation	Trigonometric ratio		Probability of distribution
Roots & real numbers	Minimum/maximum function	Truth table proof		Quartile deviation
Scalar multiplication	Monomial	Vector		Random sampling technique
Subtract radical expressions	Periodic function			Regression coefficient
Vector addition	Phase shift			Regression line

(continued)

Academic Math Vocabulary Based on NCTM Content & Process Standards Grade Band: 9–12 (continued)

Number and Operations	Algebra	Geometry	Measurement	Data Analysis & Probability	Process Standards
Vector division	Polynomial			Representativeness of sample	
Vector multiplication	Polynomial function			Sample statistic	
Vector subtraction	Polynomial solution by bisection			Sampling distribution	
	Polynomial solution by sign change			Spurious correlation	
	Polynomial solution successive approximation			Standard deviation	
	Radical expression			Statistic	
	Radical function			Statistical experiment	
	Range of function			Statistical regression	
	Rational functions			Treatment group	
	Real-world function			Two-way tables	
	Recurrence equation			Univariate distribution	
	Recurrence relationship			Variance	
	Recursive equation				
	Roots to determine cost				
	Roots to determine profits				
	Roots to determine revenue				
	Series				
	Series circuit				
	Sigma notation				

Sinusoidal function					
Speed					
Step function					
Systems of inequalities					
Term					
Trigonometric relations					
Velocity					

Triangle

One characteristic of a _____ is _____.

Square

One characteristic of a _____ is _____.

Hexagon

One characteristic of a _____ is _____.

Rhombus

One characteristic of a _____ is _____.

Trapezoid

One characteristic of a _____ is _____.

Picture	Function Rule

Written Description

Table

	Process	

Graph

Student Name & Measuring Tool	Desk	Chalkboard	Book
Hand 1. 2. 3. 4.			
Foot 1. 2. 3. 4.			
Finger 1. 2. 3. 4.			
Ruler 1. 2. 3. 4.			

The best way to measure is using a _____ because...

Vocabulary Term: _____

Teacher Definition: _____

Student Description: _____

Illustration:

An example of a nonstandard/standard unit of measurement is _____.

Measuring Tool & Student Names	Desk	Chalkboard	Book
Hand 1. 2. 3. 4.			
Foot 1. 2. 3. 4.			
Finger 1. 2. 3. 4.			
Ruler 1. 2. 3. 4.			

The best way to measure is using a _____ because...

Group Number _____

Student Names: _____

_____

Use this table below to record your measurements. In the first column you will find the classroom item to be measured. In the second, record the length of the item. In the third, record the distance between the designated items. In the fourth, record what unit you used to measure (standard and nonstandard) and in the fifth record the total number of units.

Classroom Item	Length of . . .	Distance Between . . .	Unit Used to Measure Standard or Nonstandard & Type	Number of Units
Classroom Door		Door & Teacher's Desk		
Teacher's Desk		Door & Student's Desk		
Student's Desk		Door & Window		
Classroom Window		Window & Teacher's Desk		

Word:

Triangle

Illustration of the word:

Word used in a sentence:

The Laker basketball team uses a triangle offense to score baskets.

Definition of the word:

A triangle is a geometric shape that has three sides and three angles.

Borrowed $2	Borrowed $10
Borrowed $12	Borrowed $11
Borrowed $5	Borrowed $23
Borrowed $9	Borrowed $27

Borrowed $0	**Borrowed $6**
Borrowed $30	**Borrowed $15**
Earned $2	**Earned $10**
Earned $12	**Earned $11**

Earned $5	**Earned $23**
Earned $9	**Earned $27**
Earned $0	**Earned $6**
Earned $30	**Earned $15**

I predict the room temperature of the water will be . . .

1. What is the temperature of the water at room temperature?	Glass 1	Glass 2
	The temperature of the water in glass 1 is _____ degrees C.	The temperature of the water in glass 2 is _____ degrees C.

If you add ice to the glass, I predict the temperature will . . .

If you add hot water to the glass, I predict the temperature will . . .

2. Add hot water to glass 1. What is the temperature of the water after adding hot water?	After adding hot water to glass 1, the temperature is_____degrees C.
3. By how many degrees did the temperature change?	The temperature changed by . . .

4. Add 3 ice cubes to glass 2. What is the temperature of the water after adding ice cubes?	After adding 3 ice cubes to glass 2, the temperature is _____ degrees C.
5. By how many degrees did the temperature change?	The temperature changed by . . .

6. Compare and contrast the change in temperature in glass 1 and glass 2.

**Instructions for
Who Is Colder? Card Game**

Materials: Deck of Cards

1. Students can be grouped in 2's, 3's or 4's. Give each group a deck of cards and have them remove the face cards. The Aces represent 1 and the Jokers represent 0. Red cards are negative integers and black cards are positive integers.

2. Distribute cards evenly among the players. Each student will draw one card. The students will say, "My temperature is_____degrees C." The student with the lowest temperature degree will win the cards. However, he or she must say, "I win because . . ." If there is a tie, no one wins the cards. The cards stay in the middle and the person who wins the next round gets all the cards.

3. At the end of the game, the person with the most cards wins the game.

Negative Integers

Positive Integers

<u>Winter Weather Alert</u>

Currently the temperature is -6 degrees F. It is expected for the temperature to drop another 7 degrees F. How cold will it get? What would the temperature be in C?

For the following problems, solve by using two-color counters and a number line, write an equation, and determine whether integers were added or subtracted and if the answer is positive or negative.

Problem	Two-Color Counters	Number Line	Equation	Add or Subtract	Answer + or −
1. 2 + 3					
2. −2 + (−3)					
3. 5 + (−7)					
4. −8 + 4					
5. 9 + (−2)					
6. −7 + (−3)					
7. −2 + (−7)					
8. 5 + 4					

Write real-life situations for problems 4—7.

9.	10.	11.
		12.

The temperature is 5 degrees C and it drops by 8 degrees C. What is the final temperature?	**Represent with Two-Color Counters**
Represent with Number Line	**Equation**

Problem	Two-Color Counters	Equation	Add or Subtract	Answer + or −	Apply Additive Inverse
1. −3 − 4					
2. 7 − 10					
3. −3 − (−6)					
4. −2 − 7					
5. 1 − (−5)					
6. −8 − (−4)					
7. 9 − (−2)					
8. 10 − (−5)					

Write real-life situations for problems 4—7.

9.	10.
11.	12.

Fun with Integers Instructions

Materials: Deck of cards, +/− Spinner, and Recording Sheet

1. Provide each pair of students with a deck of playing cards, spinner, and recording sheet. NOTE: Students must remove all face cards and jokers. Black aces are worth 1 and red aces are worth −1. Black numbered cards represent positive integers and red numbered cards represent negative integers.

2. Shuffle cards and place the card deck between student 1 and student 2.

3. Each student draws a card and places the value of their number on the recording sheet. One student spins the spinner and determines whether the problem will be addition or subtraction.

4. Students apply rules for adding integers and write their answer on the recording sheet.

5. Students repeat steps 2–4.

Student 1	Sign	Student 2	Equation

Spinner

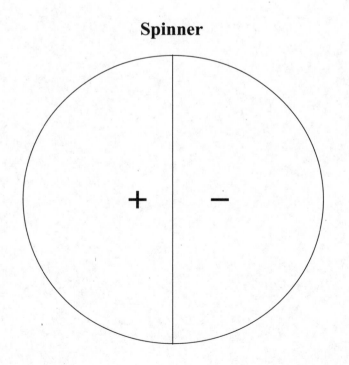

Where Is the Submarine?

A crew member of a US Submarine spots an unidentified submarine 800 feet below sea level. If the unidentified submarine ascends 250 feet, what is its new position?

Problem	Solution 1	Solution 2
1. Maria borrowed $20 from Jane and $50 from Bob. How much money did Maria borrow?		
2. An elevator is on the 10th floor. It descends 7 stories. On what floor does the elevator stop?		
3. A submarine was situated 430 feet below sea level. If it descends 200 feet, what is its new position?		
4. Mt. Everest is 29,028 feet above sea level. The Dead Sea is 1,312 feet below sea level. What is the difference between these two elevations?		

1. The squared pools below need to be tiled. You have been hired to tile the pools. The dimensions of the pools are 1ft × 1ft, 2ft × 2ft, and 3ft × 3ft. How many square tiles do you need to buy to tile each pool?
2. Fill out the table, graph the table's data, write a function rule, draw a picture, identify the domain/range and independent/dependent variables, and describe the rate of change.

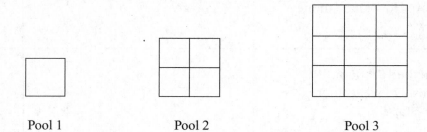

Pool 1   Pool 2   Pool 3

Picture	Function Rule

Written Description

Table	Graph

Pool Number	Process Column	Perimeter
0		
1		
2		
3		
4		
5		
x		

1. The squared gardens below need to be fenced to keep the rabbits out. You have been hired to fence the gardens. The dimensions of the gardens are 1ft × 1ft, 2ft × 2ft, and 3ft × 3ft. How many feet of fence do you need to buy to fence each garden?
2. Fill out the table, graph the table's data, write a function rule, draw a picture, identify the domain/range and independent/dependent variables, and describe the rate of change.

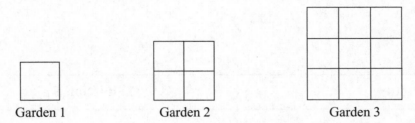

Garden 1   Garden 2   Garden 3

Picture	Function Rule

Written Description

Table			Graph

Garden Number	Process Column	Perimeter
0		
1		
2		
3		
4		
5		
x		

1. Grab a sheet of paper. How many layers of paper are there? Fold the paper in half. How many layers of paper are there? Record your findings on the table below.
2. Continue folding and fill out the table, graph the table's data, write a function rule, draw a picture, identify the domain/range and independent/dependent variables, and describe the rate of change.

Picture	Function Rule

Written Description

Table

Number of Folds	Process Column	Number of Layers
0		
1		
2		
3		
4		
5		
x		

Graph

	Tiling Squared Pools	Fencing Gardens	Layers Galore	
Type of Function				**Similarities & Differences**
Function Rule				**Similarities & Differences**
What is the domain and range?				**Similarities & Differences**
Is the function discrete or continuous?				**Similarities & Differences**
Rate of Change				**Similarities & Differences**
Describe table patterns				**Similarities & Differences**
Describe the graph				**Similarities & Differences**

Use a graphing calculator to determine how the parent function will change.

Parent Function	Graph of Parent Function	Equation	Table	Graph	What happens to the graph of a function when you add a constant to the parent function?
$y = x$		$y = x + 3$	X Y		What happens to the graph is . . .
$y = x^2$		$y = x^2 - 1$	X Y		What happens to the graph is . . .
$y = 2^x$		$y = 2^x - 2$	X Y		What happens to the graph is . . .

Use a graphing calculator to determine how the parent function will change.

Parent Function	Graph of Parent Function	Equation	Table	Graph	What happens to the graph of a function when you replace x with $(x-a)$?
$y = x$		$y = (x-3)$	X Y		What happens to the graph is . . .
$y = x^2$		$y = (x+1)^2$	X Y		What happens to the graph is . . .
$y = b^x$		$y = 3^{(x-1)}$	X Y		What happens to the graph is . . .

Use a graphing calculator to determine how the parent function will change.

Parent Function	Graph of Parent Function	Equation	Table	Graph	What happens to the graph of a function when you replace x with (x−a) and add a constant to the parent function?
$y = x$		$y = (x - 3) + 2$	X Y		What happens to the graph is . . . .
$y = x^2$		$y = (x - 4)^2 + 2$	X Y		What happens to the graph is . . . .
$y = b^x$		$y = 2^{(x-1)} + 2$	X Y		What happens to the graph is . . . .

(0, 0)	**(-1, 1)**
(1, 1)	**(2, 4)**
(-2, 4)	

(0, 0)	**(-1, -1)**
(1, 1)	**(-2, -2)**
(2, 2)	

(0, 1)	**(-1, 0.5)**
(1, 2)	**(2, 4)**
(-2, 0.25)	

$y = x - 1$	$y = x + 40$	$y = x$
$y = x - 11$	$y = x - 5$	$y = (x - 1) + 2$
$y = x + 5$	$y = (x - 3) - 11$	$y = (x + 2) - 11$

$y = x^2 - 1$	$y = x^2 + 40$	$y = x^2$
$y = x^2 - 11$	$y = x^2 - 5$	$y = (x-1)^2 + 2$
$y = x^2 + 5$	$y = (x-3)^2 - 11$	$y = (x+2)^2 - 11$

$y = 2^x - 1$	$y = 3^x + 40$	$y = 2^x$
$y = 2^x - 11$	$y = 2^x - 5$	$y = 3^x + 2$
$y = 2^x + 5$	$y = 3^x - 11$	$y = 3^x - 11$

Use a graphing calculator to determine how the parent function will change.

Parent Function	Graph of Parent Function	Equation	Table	Graph	How did the graph of the parent function change?
y		$y = -\dfrac{1}{2}x$	X Y		The graph of the parent function changed by . . .
$y = x^2$		$y = -\dfrac{2}{3}x^2$	X Y		The graph of the parent function changed by . . .
$y = 2^x$		$y = 2^{(-0.4x)}$	X Y		The graph of the parent function changed by . . .

Use a graphing calculator to determine how the parent function will change.

Parent Function	Graph of Parent Function	Equation	Table	Graph	How did the graph of the parent function change?
$y = x$		$y = \dfrac{1}{2}x$	X Y		The graph of the parent function changed by . . .
$y = x^2$		$y = \dfrac{2}{3}x^2$	X Y		The graph of the parent function changed by . . .
$y = b^x$		$y = 2^{(0.4x)}$	X Y		The graph of the parent function changed by . . .

Use a graphing calculator to determine how the parent function will change.

Parent Function	Graph of Parent Function	Equation	Table	Graph	How did the graph of the parent function change?
$y = x$		$y = 3x$	X Y		The graph of the parent function changed by . . .
$y = x^2$		$y = 4x^2$	X Y		The graph of the parent function changed by . . .
$y = 2^x$		$y = 2^{(4x)}$	X Y		The graph of the parent function changed by . . .

33 multiplying x by a<−1 lab sheet

Use a graphing calculator to determine how the parent function will change.

Parent Function	Graph of Parent Function	Equation	Table	Graph	How did the graph of the parent function change?
$y = x$		$y = -5x$	X Y		The graph of the parent function changed by . . . .
$y = x^2$		$y = -3x^2$	X Y		The graph of the parent function changed by . . . .
$y = b^x$		$y = 2^{(-4x)}$	X Y		The graph of the parent function changed by . . . .

Use a graphing calculator to determine how the parent function will change.

Parent Function	Graph of Parent Function	Equation	Table	Graph	How did the graph of the parent function change?
$y = x$		$y = -x$	X Y		The graph of the parent function changed by . . . .
$y = x^2$		$y = -x^2$	X Y		The graph of the parent function changed by . . . .
$y = b^x$		$y = 2^{(-x)}$	X Y		The graph of the parent function changed by . . . .

35 combining transformations lab sheet

Use a graphing calculator to determine how the parent function will change.

Parent Function	Graph of Parent Function	Equation	Table	Graph	How did the graph of the parent function change?
$y = x$		$y = -3(x-3)+2$	X Y		The graph of the parent function changed by . . .
$y = x^2$		$y = -\frac{1}{2}(x-4)^2+2$	X Y		The graph of the parent function changed by . . .
$y = b^x$		$y = 6(2^{(2x-1)})+2$	X Y		The graph of the parent function changed by . . .

references

American Educational Research Association (AERA), American Psychological Association (APA), & National Council on Measurement in Education (NCME). (2000). Position statement of the American Educational Research Association concerning high-stakes testing in pre-K-12 education. *Educational Researcher, 29*, 24–25.

August, D., & Shanahan, T. (Eds.). (2006). *Developing literacy in second-language learners: A report of the National Literacy Panel on Language-Minority Children and Youth*. Mahwah, NJ: Erlbaum.

Aukerman, M. (2007). A culpable CALP: Rethinking the conversational/academic proficiency distinction in early literacy instruction. *The Reading Teacher, 60*(7), 626–635.

Bailey, A. L. (Ed.). (2007). *The language demands of school: Putting academic English to the test*. New Haven, CT: Yale University Press.

Baumann, J., Jones, L., & Seifert-Kessell, N. (1993). Using think-alouds to enhance children's comprehension monitoring abilities. *The Reading Teacher, 47*(3), 184–193.

Biancarosa, G., & Snow, C. (2004). *Reading next: A vision for action and research in middle and high school literacy*. Report to the Carnegie Corporation of New York. Washington, DC: Alliance for Excellent Education.

Buehl, D. (2009). *Classroom strategies for interactive learning*. Newark, DE: International Reading Association.

Castillo, M. (2008). Reviewing objectives with English language learners. Presented at SEI Seminar, Phoenix, AZ.

Cazden, C. (1976). How knowledge about language helps the classroom teacher—or does it? A personal account. *The Urban Review, 9*, 74–91.

Cazden, C. (1986). Classroom discourse. In M. D. Wittrock (Ed.), *Handbook of research on teaching* (3rd ed. pp. 432–463). New York: Macmillan.

Cazden, C. (2001). *Classroom discourse: The language of teaching and learning, second edition*. Portsmouth, NH: Heinemann.

Center for Applied Linguistics. (2007). *Academic literacy through sheltered instruction for secondary English language learners*. Final Report to the Carnegie Corporation of New York. Washington, DC: Center for Applied Linguistics.

Chamot, A. U., & O'Malley, J. M. (1994). The CALLA handbook: Implementing the cognitive academic language learning approach. Reading, MA: Addison-Wesley.

Collier, V. (1987). Age and rate of acquisition of second language for academic purposes. *TESOL Quarterly, 21*(3), 617–641.

Cosentino de Cohen, C., Deterding, N., & Clewell, B.C. (2005). *Who's left behind: Immigrant children in high and low LEP schools*. Washington, DC: Urban Institute. Retrieved January 2, 2009 at http://www.urban.org/UploadedPDF/411231_whos_left_behind.pdf

Coxhead, A. (2000). A new academic word list. *TESOL Quarterly, 34*(2), 213–238.

Cuevas, G. (2005). Teaching mathematics to English language learners: Perspectives for effective instruction. In Huerta-Macías, A. G., *Working with English language learners: Perspectives and practice* (pp. 69–86). Dubuque, IA: Kendall-Hunt.

Cummins, J. (1979). *Cognitive/academic language proficiency, linguistic interdependence, the optimum age questions, and some other matters*. Working Papers on Bilingualism, No. 19, 121–129. Toronto: Ontario Institute for Studies in Education.

Cummins, J. (2000). *Language, power, and pedagogy: Bilingual children in the crossfire.* Clevedon, UK: Multilingual Matters.

Cummins, J. (2006). How long does it take for an English language learner to become proficient in a second language? In E. Hamayan & R. Freeman (Eds.), *English language learners at school: A guide for administrators* (pp. 59–61). Philadelphia: Caslon Publishing.

Dutro, S., & Moran, C. (2003) Rethinking English language instruction: An architectural approach. In G. Garcia (Ed.), *English learners: Reaching the highest level of English literacy* (pp. 227–258). Newark, NJ: International Reading Association.

Echevarria, J. (1995). Interactive reading instruction: A comparison of proximal and distal effects of instructional conversations. *Exceptional Children, 61*(6), 536–552.

Echevarria, J., & Graves, A. (in press). *Sheltered content instruction: Teaching English language learners with diverse abilities* (4th ed.). Boston: Allyn & Bacon.

Echevarria, J., Richards, C., Canges, R. & Francis, D. (in review). The role of language in the acquisition of science concepts with English learners. *Journal of Research on Educational Effectiveness*.

Echevarria, J., & Short, D. (2009). Programs and practices for effective sheltered content instruction. In D. Dolson & L. Burnham-Massey (Eds.), *Improving education for English learners: Research-based approaches*. Sacramento, CA: California Department of Education.

Echevarria, J., Short, D., & Powers, K. (2006). School reform and standards-based education: An instructional model for English language learners. *Journal of Educational Research, 99*(4), 195–211.

Echevarria, J., Short, D., & Vogt, M.E. (2008). *Implementing the SIOP® model through effective professional development and coaching.* Boston, MA: Pearson/Allyn & Bacon.

Echevarria, J., & Silver, J. (1995). *Instructional conversations: Understanding through discussion*. [Videotape]. National Center for Research on Cultural Diversity and Second Language Learning.

Echevarria, J., Vogt, M.E., & Short, D. (2004). *Making content comprehensible for English learners: The SIOP® Model* (2nd ed.). Boston: Allyn & Bacon.

Echevarria, J., Vogt, M.E., & Short, D. (2008). *Making content comprehensible for English learners: The SIOP® Model* (3rd ed.). Boston: Allyn & Bacon.

Echevarria, J., Vogt, M.E., & Short, D. (2010a). *Making content comprehensible for elementary English learners: The SIOP® Model*. Boston: Allyn & Bacon.

Echevarria, J., Vogt, M.E., & Short, D. (2010b). *Making content comprehensible for secondary English learners: The SIOP® Model*. Boston: Allyn & Bacon.

Fisher, D., & Frey, N. (2008). *Better learning through structured teaching*. Alexandria, VA: Association for Supervision and Curriculum Development.

Flynt, E. S., & Brozo, W. G. (2008). Developing academic language: Got words? *The Reading Teacher, 61*(6), 500–502.

Gall, M. D. (1984). Synthesis of research on questioning. *Educational Leadership, 42,* 40–47.

Gandara, P., Maxwell-Jolly, J., & Driscoll, A. (2005). *Listening to teachers of English language learners: A survey of California teachers' challenges, experiences, and professional development needs*. Santa Cruz, CA: The Center for the Future of Teaching and Learning.

Garcia, G., & Beltran, D. (2003). Revisioning the blueprint: Building for the academic success of English learners. In G. Garcia (Ed.), *English learners: Reaching the highest levels of English literacy*. Newark, DE: International Reading Association.

Garcia, G. E., & Godina, H. (2004). Addressing the literacy needs of adolescent English language learners. In T. Jetton & J. Dole (Eds.), *Adolescent literacy: Research and practice* (pp. 304–320). New York: The Guildford Press.

Gersten, R., Baker, S. K., Shanahan, T., Linan-Thompson, S., Collins, P., & Scarcella, R. (2007). *Effective literacy and English language instruction for English learners in the Elementary grades: A practice guide* (NCEE 2007–4011). Washington, DC: National Center for Education Evaluation and Regional Assistance, Institute of Education Sciences, U.S. Department of Education. Retrieved from http://ies.ed.gov/ncee.

Genesee, F., Lindholm-Leary, K., Saunders, W., & Christian, D. (2006). *Educating English language learners: A synthesis of research evidence*. New York: Cambridge University Press.

Goldenberg, C. (2008). Teaching English language learners: What the research does—and does not—say. *The American Educator, 32*(2), 8–23.

Graham, S., & Perin, D. (2007). *Writing next: Effective strategies to improve writing of adolescents in middle and high schools*. A report to the Carnegie Corporation of New York. Washington, DC: Alliance for Excellent Education.

Hayden, D., & Cuevas, G. (1990). *Pre-Algebra lexicon*. Washington, DC: Center for Applied Linguistics.

Hiebert, E. H. (2005). *Word Zones™: 5,586 most frequent words in written English*. Available at www.textproject.org.

Hiebert, E. H. (2005). *1,000 most frequent words in middle-grades and high school texts*. Available at www.textproject.org.

Kindler, A. (2002). *Survey of the states' limited English proficient students and available educational programs and services. 2000–01 summary report*. Washington, DC: National Clearinghouse for English Language Acquisition.

Kober, N., Zabala, D., Chudowsky, N., Chudowsky, V., Gayler, K., & McMurrer, J. (2006). *State high school exit exams: A challenging year*. Washington, DC: Center on Education Policy.

Krashen, S. (1985). *The input hypothesis: Issues and implications*. London: Longman.

Lee, J., Grigg, W., & Dion, P. (2007). *The nation's report card: Mathematics 2007*. (NCES 2007–494). U.S. Department of Education, Institute of Education Sciences, National Center for Education Statistics. Washington, DC: U.S. Government Printing Office.

Lee, J., Grigg, W., & Donahue, P. (2007). *The nation's report card: Reading 2007*. (NCES 2007–496). U.S. Department of Education, Institute of Education Sciences, National Center for Education Statistics. Washington, DC: U.S. Government Printing Office.

Lee, O. (2005). Science education with English language learners: Synthesis and research agenda. *Review of Educational Research, 75*(4), 491–530.

Mehan, H. (1979). *Learning lessons*. Cambridge: Harvard University Press.

National Council of Teachers of Mathematics (NCTM). *Principles and standards for school mathematics*. Reston, VA. Retrieved May 14, 2009 at http://www.nctm.org/standards/default.aspx?id=58.

National Institute of Child Health and Human Development (NICHD). (2000). Report of the National Reading Panel. *Teaching Children to read: An evidence-based assessment of the scientific literature on reading and its implications for reading instruction*. (NIH Publication No. 00-4769). Washington, DC: U.S. Department of Health and Human Services.

Oczkus, L. (2009). *Interactive think-aloud lessons: 25 surefire ways to engage students and improve comprehension*. New York: Scholastic and Newark, DE: International Reading Association.

Ogle, D. (1986). K-W-L: A teaching model that develops active reading of expository text. *The Reading Teacher, 39,* 564–570.

Reiss, J. (2008). *102 content strategies for English language learners*. Upper Saddle River, NJ: Pearson/Merrill Prentice Hall.

Saunders, W., & Goldenberg, C. (1992). *The effects of instructional conversations on transition students' concept development*. Revised version of paper presented at the annual meeting of the American Educational Research Association, San Francisco, CA.

Saunders, W., & Goldenberg, C. (2007). The effects of an instructional conversation on English Language Learners' concepts of friendship and story comprehension. In R. Horowitz (Ed.), *Talking texts: How speech and writing interact in school learning* (pp. 221–252). Mahwah, NJ: Erlbaum.

Saunders, W., & Goldenberg, C. (in press). Research to guide English language development. In D. Dolson & L. Burnham-Massey (Eds) *Improving Education for English Learners: Research-Based Approaches*. Sacramento, CA: California Department of Education.

Scott, J. A., Jamison-Noel, D., & Asselin, M. (2003). Vocabulary instruction the day in twenty-three Canadian upper elementary classrooms. *The Elementary School Journal, 103*, 269–286.

Short, D., & Richards, C. (2008). *Linking science and academic English: Teacher development and student achievement.* Paper presented at the Center for Research on the Educational Achievement and Teaching of English Language Learners (CREATE) Conference, Minneapolis, MN, October, 2008.

Short, D., Vogt, M.E., & Echevarria, J. (in press). *The SIOP® Model for Teaching Science to English Learners.* Boston: Allyn & Bacon.

Short, D., Vogt, M.E., & Echevarria, J. (in press). *The SIOP® Model for Teaching History-Social Studies to English Learners.* Boston: Allyn & Bacon.

Stahl, S. A., & Nagy, W. E. (2006). *Teaching word meanings.* Mahwah, NJ: Erlbaum.

Suarez-Orozco, C., Suarez-Orozco, M. M., & Todorova, I. (2008). *Learning in a new land: Immigrant students in American society.* Cambridge, MA: Harvard University Press.

Tharp, R., & Gallimore, R. (1988). *Rousing minds to life: Teaching, learning and schooling in social context.* Cambridge: Cambridge University Press.

U.S. Department of Education. (2006). *Building partnerships to help English language learners.* Fact sheet. Retrieved January 2, 2008 at http://www.ed.gov/nclb/methods/english/lepfactsheet.html.

Vogt, M.E., & Echevarria, J. (2008). *99 ideas and activities for teaching English learners with the SIOP® Model.* Boston: Allyn & Bacon.

Walqui, A. (2006). Scaffolding instruction for English language learners: A conceptual framework. *The International Journal of Bilingual Education and Bilingualism, 9*(2), 159–180.

Vogt, M.E., & Echevarria, J., & Short, D. (in press). *The SIOP® Model for Teaching English-Language Arts to English Learners.* Boston: Allyn & Bacon.

Watson, K., & Young, B. (1986). Discourse for learning in the classroom. *Language Arts, 63*(2), 126–133.

Zehler, A. M., Fleishman, H. L., Hopstock, P. J., Stephenson, T. G., Pendzik, M. L., & Sapru, S. (2003). *Descriptive study of services to LEP students and to LEP students with disabilities; Policy report: Summary of findings related to LEP and SpEd-LEP students.* Arlington, VA: Development Associates.

Zwiers, J. (2004). *Developing academic thinking skills in grades 6–12.* Newark, DE: International Reading Association.

Zwiers, J. (2008). *Building academic language: Essential practices for content classrooms.* San Francisco: Jossey-Bass; Newark, DE: International Reading Association.

index